# TOPICS IN CLIMATE MODELING

Edited by **Theodore Hromadka**
and **Prasada Rao**

## Topics in Climate Modeling

http://dx.doi.org/10.5772/61591
Edited by Theodore Hromadka and Prasada Rao

### Contributors

Narges Khanjani, Sergei Soldatenko, Denis Chichkine, Tadesse Zeleke, Fentahun Agidew, Ted Hromadka, James Williams, Britton Landry, Matthew Mogensen, Dong Jiang, Jianhua Wang, Yaohuan Huang, Xiaoyang Song, Fedor Mesinger, Katarina Veljovic, Sin Chan Chou, Jorge Luis Gomes, Andre Lyra

**CBS,** Edition **2017**

**Published by InTech**

Janeza Trdine 9, 51000 Rijeka, Croatia



**Publishing Process Manager** Ana Pantar

**Technical Editor** SPi Global

**Cover Designer** InTech Design Team

Additional hard copies can be obtained from orders@intechopen.com

Topics in Climate Modeling, Edited by Theodore Hromadka and Prasada Rao
p. cm.
Print ISBN 978-953-51-2660-7
Online ISBN 978-953-51-2661-4

INTECH

open science | open minds

# Contents

# Preface

The topics of climate change, weather prediction, atmospheric sciences and other related fields are gaining increased attention due to the possible impacts of changes in climate and weather upon the planet. Concurrently, the increasing ability to computationally model the governing partial differential equations that describe these various topics of climate has gained a great deal of attention as well. These topics are focused upon changes to the planet and continue to evolve as data bases are enlarged with new large scale studies, of which the ability to share such data on a global scale has greatly expanded. The ability to computationally model these evolving changes and predict possible outcomes from these changes continue to advance at a rapid scale with enhancements in computational engineering mathematics, including the mainstream theme of computational fluid dynamics.

This book will examine several of the topics that are used to model the climate and predict changes as a stepping stone to recent advances in the fields of study and provide focal points of endeavor in the evolving technology.

Several chapters provide a summary of key topics currently of high interest to the climate modeling community. In the first chapter the authors present the vector analysis mathematical theory behind compressible and incompressible flows. The second chapter presents the results of a regional climate model (RegCM4) model for upper blue Nile river basin. The third chapter presents the time-series analysis of water energy balance coupled with ecosystem variability in South West China. The fourth chapter reviews the effect of climate change on human health. The fifth chapter quantifies and details the sensitivity of numerical solution to different input parameters. The sixth chapter reviews the evolution of Eta model and underscores its potential to simulate complex climate change patterns.

**Dr. Theodore Hromadka**
Professor, Department of Mathematical Sciences, United States Military Academy,
West Point, NY, USA

Research Professor,
Department of Civil and Environmental Engineering,
California State University,
Fullerton, CA, USA

**Dr. Prasada Rao**
Professor, Department of Civil and Environmental Engineering,
California State University,
Fullerton, CA, USA

Chapter 1

# Computational Vector Mechanics in Atmospheric and Climate Modeling

James Williams, Britton Landry,
Matthew Mogensen and T. V. Hromadka II

Additional information is available at the end of the chapter

http://dx.doi.org/10.5772/65009

**Abstract**

The mathematical underpinnings of vector analysis are reviewed as they are applied in the development of the ensemble of numeric statements for subsequent matrix solution. With the continued advances in computational power, there is increased interest in the field of atmospheric modelling to decrease the computational scale to a micro-scale. This interest is partially motivated by the ability to solve large scale matrix systems in the number of occasions to enable a small-scale time advancement to be approximated in a finite-difference scheme. Solving entire large scale matrix systems several times a modelling second is now computationally feasible. Hence the motivation to increase computational detail by reducing modelling scale.

**Keywords:** Vectors, vector fields, curl, divergence, Divergence Theorem

## 1. Introduction

Computational Engineering Mathematics as applied to topics in Computational Geosciences and Computational Fluid Dynamics, among other themes and topics, is the foundation of computational modelling processes involved with Climate and Atmospheric processes. This chapter reviews the key mathematical underpinnings of computational mathematics used to represent climate, atmospheric and hydrologic related processes. The commonly used numerical and computational approaches of finite difference, finite element, and finite volume have common roots stemming from the mathematical analysis of partial differential equations, ordinary differential equations, vector differential and integral equations, among other topics including solution of large scale matrix systems involving tens of thousands of unknown variables, time-stepped numerous times a modelling second, resulting in incredible numbers of computations. This chapter focuses on important aspects of matrix solutions and on the vector analysis towards use in development of finite volume and nodal domain numeric solutions of the governing Navier-Stokes equations. Due to the continuing increasing computational capability forecast according to Moore's Law, the use of Navier-Stokes solvers in routine problems is becoming more prevalent, and will possibly be common practice in a few short years.

Methods for developing large scale matrix systems and their solution is a fundamental area of knowledge needed for the successful modelling of atmospheric processes and climate. This particular topic can be investigated in other references focused towards solving large scale matrix systems. Methods for efficiently constructing the matrix system for subsequent solution is also a subject for other references. Rather, in the current chapter, the focus is on the topics of matrix system solution and the existence and uniqueness of solutions. The reduction of the governing partial differential equations into numeric form or into finite element or finite volume form, and then further developed into numeric statements, is another subject dealt with in detail in other references.

In the current chapter, the mathematical underpinnings of vector analysis are reviewed as they are applied in the development of the ensemble of numeric statements for subsequent matrix solution. With the continued advances in computational power, there is increased interest in the field of atmospheric modelling to decrease the computational scale to a micro-scale. This interest is partially motivated by the ability to solve large scale matrix systems in the number of occasions to enable a small-scale time advancement to be approximated in a finite-difference scheme. Solving entire large scale matrix systems several times a modelling second is now computationally feasible. Hence the motivation to increase computational detail by reducing modelling scale.

## 2. The Navier-Stokes equations in computational engineering mathematics

The mathematical description of fluid processes are embodied in the well-known Navier-Stokes equations. These equations relate Newton's second law of motion to a control volume of fluid. Their origins can be found in the work dated between 1827 and 1845, including the advances made by Saint-Venant. Although all fluids possess viscous effects, in many problem situations the effects of viscosity can be neglected, reducing the governing equations considerably. The assumption of the fluid being incompressible in the range of temperatures and conditions under study further reduces the governing equations to being applicable to a fluid that is incompressible and having zero viscosity. That is, an ideal fluid. The analysis of ideal fluid flow is considerably simpler to undertake than the full Navier-Stokes equations, and provides generally a good framework and basis for the analysis of the complete fluid flow mathematical regime. Ideal fluid flow ignores shear stresses in fluids and normal stresses (that is, thermodynamic pressure forces) are the internal force components carried forward into the analysis.

http://dx.doi.org/10.5772/65009

The resulting Euler equations are well known in fluid mechanics and are particularly amenable to solution when applied to flow on a streamline. For steady flow, the Euler equations indicate that flow velocity and pressure sum to as constant value along a horizontal streamline (or neglecting weight of the fluid control volume). Integration of the Euler equation along a streamline for steady flow conditions results in the Bernoulli equation, which is a commonly used relationship in the analysis of fluid flow. The Bernoulli equation is often misapplied by failure to adhere to the assumptions applied in the derivation of that equation. Namely, steady flow conditions, frictionless or shear-free flow, incompressible fluid flow, and flow on a streamline. A common misapplication is to apply Bernoulli's equation at two points within a fluid domain that do not lie on the same streamline. The sum of the Bernoulli equation components is called the Bernoulli constant, and has different values depending on the streamline under study. If the Bernoulli constant is constant throughout the entire fluid domain, regardless of streamline, then the flow is said to be irrotational. Irrotational flow can be assessed within a vector field by use of the vector operator Curl, defined in the development below. In particular, the streamline can be mathematically described in terms of vector notation with respect to usual measure of distance along the streamline, $s$.

When dealing with fluid flow, particularly atmospheric fluid flow, thermodynamic pressure is a key component. Thermodynamic pressure is also called static pressure. The stagnation pressure is measured under an ideal frictionless process where the fluid is decelerated to zero speed (note that velocity is a vector whereas speed is a scalar term.) The dynamic pressure quantifies the magnitude of the flow velocity. The flow speed is determined given the stagnation pressure and the static pressure at a point. These relationships become impacted by the speed, particularly with higher Mach numbers greater than about 0.3.

Application of the Bernoulli equation finds good use in many problems of high interest involving fluid flow regimes involving contraction or expansion of flows. In all of these applications, however, care is required to preserve the fundamental assumptions employed in the derivation of the Bernoulli equation.

The above discussion focuses on the mass continuity and flow momentum equations embodied in the Navier-Stokes equations. A similar resulting formulation arises using the energy equation. In this second situation, a control volume is determined that conforms to the stream lines of the flow regime, and steady incompressible flow is considered that is inviscid. The Bernoulli equation then arrives even though different concepts and boundary conditions are employed. Therefore, the Laws of Thermodynamics for steady, incompressible flow along a streamline simplifies to the well-known Bernoulli equation which is also employed for analysis of mechanical energy fluid flow problems.

In civil engineering hydraulics problems, the energy equation formulation of the Bernoulli equation is typically shown graphically, as the Energy Grade Line or EGL displaying the sum of the three energy head components of pressure head plus velocity head plus gravity head, all in units of length. The sum of these three components is the Total Energy Head, which represents the height the fluid would rise up to in a tube open for the fluid to access. The Hydraulic Grade Line or HGL is plotted the velocity head below the EGL.

## 3. Irrotational flow and the vector operator curl

Another important vector calculus operator is divergence. This operator provides several measures of fluid flow and fluid properties including compressibility, and also a measure of the strength of source or sink conditions. A divergence value that is positive indicates outwards flux from the target point, whereas a negative value indicates flow trends indicative of a sink.

The vector operators of curl and divergence are key vector calculus tools used in the analysis of fluid flows of both compressible and incompressible flow regimes, and form the basis of continuity equations.

Modeling topics currently tend to divide along lines of numerical methods, particularly in leveraging parallelism versus increased processor performance. The partial differential equations describing the atmospheric processes involved still are heavily influenced by processes such as rainfall, convective processes, and others. Many models use grid-based finite-difference analogs, whereas other approaches are based upon finite element and finite volume schemes. Because many of these schemes involves grid density problems at the poles, further attention has been paid towards use of discretization's of the globe into different geometric finite volumes and tessellations. Finite volume methods, like finite element models, are attractive in conserving energy, mass, and momentum. Furthermore, the Divergence Theorem has a more direct application with use of finite volume or finite element methods.

In this chapter, the fundamental vector calculus principles are reviewed that are most relevant to fluid flow modeling such as involved with compressible and incompressible flow problems in water and the atmosphere. Details regarding these fields far surpass the scope of this chapter, but the fundamental elements are generally based in the computational mathematics involved in formulation of and solution of large scale matrix equations.

## 4. Vectors

### 4.1. Definition and representation

A vector is used to describe a quantity such as displacement, velocity, or force that has both magnitude and direction. In two dimensions or three dimensions, we can represent the magnitude and direction of a vector with an arrow (**Figure 1**) [1].

By convention, a vector is usually written in bold, or with an arrow, such as **v** or $\vec{v}$. We may also represent a vector as a displacement vector. For example, the displacement from point A to point B can be written as the vector $\overrightarrow{AB}$ (**Figure 2**).

It is important to note that a vector is not anchored to any particular place in the coordinate system. Rather, it represents only a *change* in terms of distance and direction. Thus, two vectors are equal if they have the same magnitude and direction (**Figure 3**).

http://dx.doi.org/10.5772/65009

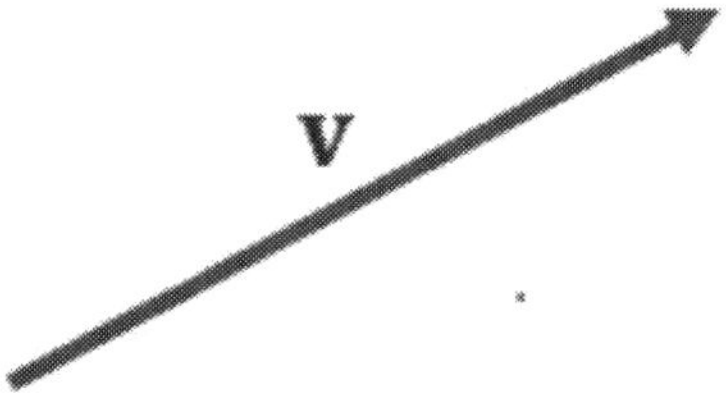


**Figure 1.** Vector representation.

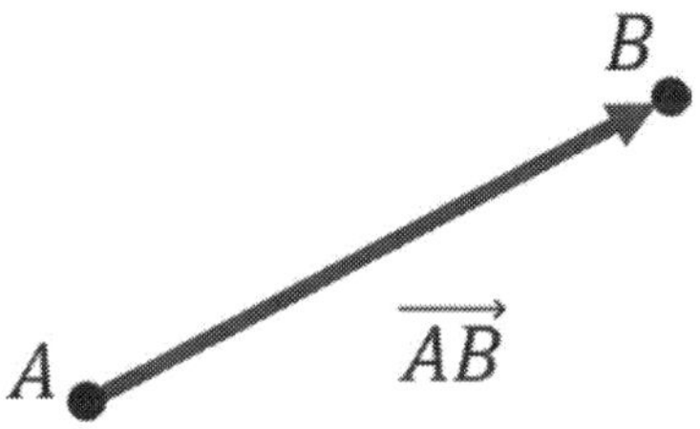


**Figure 2.** A displacement vector.

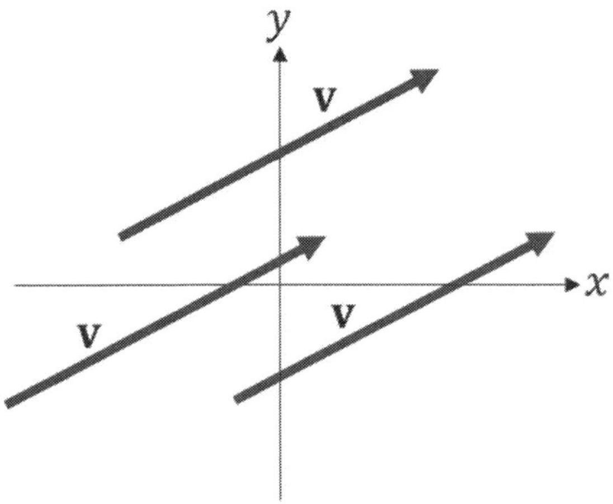


**Figure 3.** The same vector shown three times.

Most often when dealing with vectors of more than two components, it is easier to represent them algebraically. We write a vector as a list of its components, which are equivalent to the change in each of the coordinate directions (**Figure 4**). This is fairly easy to represent in two or even three dimensions, but also extends to $n$ dimensions.

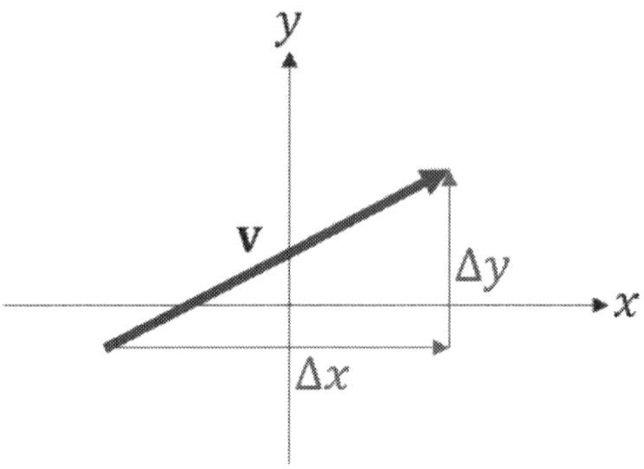


**Figure 4.** A vector $< \Delta x, \quad \Delta y >$ with illustrated components.

A vector that denotes a specific point in the coordinate system is called a position vector (meaning a vector that, when its tail is placed at the origin, points to a specific point) Thus, a position vector for a point $(a, b)$ woud be $< a, b >$. (**Figure 5**)

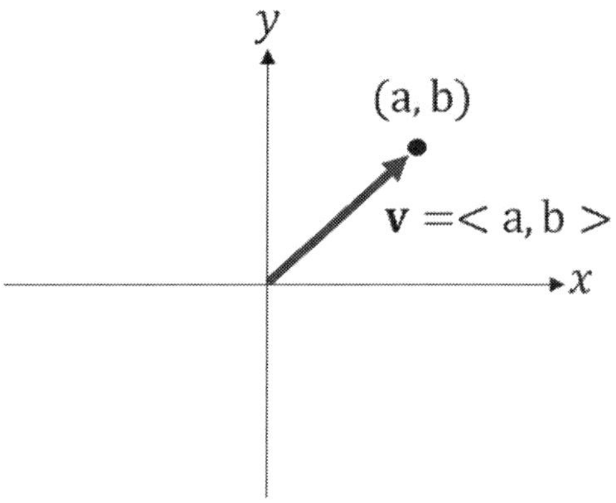


**Figure 5.** A position vector.

### 4.2. Vector addition and scalar multiplication

Two basic operations that can be performed on vectors are addition and scalar multiplication. In other words, we can add and subtract vectors, as well as multiply them by a constant which will lengthen or shorten the vector.

**Definition:** If **u** and **v** are two vectors positioned so the initial point of **v** is at the terminal point of **u**, then the **sum u+v** is the vector from the initial point of **u** to the terminal point of **v** [stewart8]. (**Figure 6**).

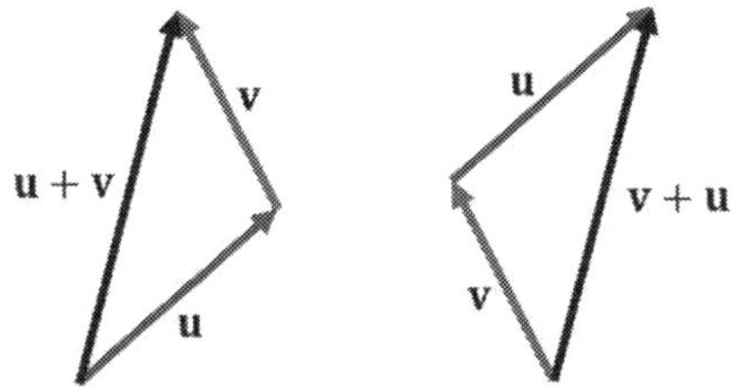


**Figure 6.** Vector addition [1].

Algebraically, vector addition results in adding the vectors component-wise:

$$\begin{aligned}&\text{Let } \mathbf{u} = \langle a,b\rangle \text{ and } \mathbf{v} = \langle c,d\rangle\\&\mathbf{u}+\mathbf{v} = \langle a,b\rangle + \langle c,d\rangle = \langle a+c, b+d\rangle\end{aligned} \tag{1}$$

In addition to representing a vector by its components, we can represent a vector by a linear combination of the basis vectors, a basis vector being a vector with a magnitude of one in the direction of only one of the coordinate axes.

$$\mathbf{r} = \langle a,b,c\rangle = a\mathbf{i} + b\mathbf{j} + c\mathbf{k} \tag{2}$$

Vector addition results in factoring each basis vector:

$$\begin{aligned}&\text{Let } \mathbf{u} = a\mathbf{i} + b\mathbf{j} \text{ and } \mathbf{v} = c\mathbf{i} + d\mathbf{j}\\&\mathbf{u}+\mathbf{v} = a\mathbf{i} + b\mathbf{j} + c\mathbf{i} + d\mathbf{j} = (a+c)\mathbf{i} + (b+d)\mathbf{j}\end{aligned} \tag{3}$$

**Definition:** If $c$ is a scalar and $\mathbf{v}$ is a vector, then the scalar multiple $c\mathbf{v}$ is the vector whose length is $|c|$ times the length of $\mathbf{v}$ and whose direction is the same as $\mathbf{v}$ if $c > 0$ and is opposite to $\mathbf{v}$ if $c < 0$. If $c = 0$ or $\mathbf{v} = 0$, then $c\mathbf{v} = 0$ [1].

Algebraically, scalar multiplication results in multiplying each component of the vector by the scalar constant:

$$\begin{aligned}&\text{Let } \mathbf{u} = \langle a,b\rangle \text{ and } c \text{ be a scalar}\\&c\mathbf{u} = \langle ca + cb\rangle = (ca)\mathbf{i} + (cb)\mathbf{j}\end{aligned} \tag{4}$$

Vector subtraction results from applying both vector addition and scalar multiplication by $-1$. Furthermore, from this we can derive from geometric or algebraic arguments the following properties of vectors:

Given **a**, **b**, **c** are vectors and $c$, $d$ are scalars: [1]

$$\mathbf{a}+\mathbf{b}=\mathbf{b}+\mathbf{a} \tag{5}$$

(commutative property)

$$\mathbf{a}+(\mathbf{b}+\mathbf{c})=(\mathbf{a}+\mathbf{b})+\mathbf{c} \tag{6}$$

$$\mathbf{a}+(-\mathbf{a})=0 \tag{7}$$

$$(c+d)\mathbf{a}=c\mathbf{a}+d\mathbf{a} \tag{8}$$

$$c(\mathbf{a}+\mathbf{b})=c\mathbf{a}+c\mathbf{b} \tag{9}$$

The length of a vector can be found by utilizing the Pythagorean Theorem. If a vector has the components $\mathbf{u} = \ < a, b, c >$, then the length of **u** is found by: $|\mathbf{u}| = \sqrt{a^2 + b^2 + c^2}$

### 4.3. The dot product

Multiplication between vectors is not the same as between scalars. With vectors we have two "products", the dot product and the cross product. The dot product between vectors will play a vital part in vector calculus. We can think of the dot product as being a metric of how much one vector is influencing another. An application for the dot product is work, where work is defined as $work = Force * Distance$. If we think of the force and distance as vectors, then the amount of work to move a sled between two points as one pulls on a rope is the dot product between the force and direction or distance vectors (**Figure 7**).

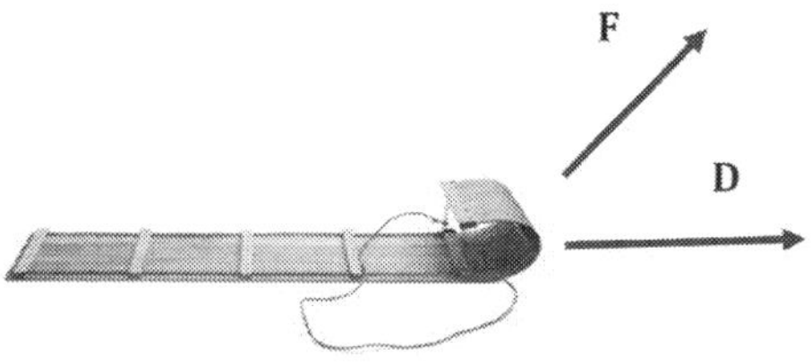


**Figure 7.** Dot product visualization. Image courtesy www.newyorkfamily.com.

**Definition:** If $\mathbf{a} = \ < a_1, a_2, ..., a_n >$ and $\mathbf{b} = \ < b_1, b_2, ..., b_n >$ are $n$ dimensional vectors, then the dot product of **a** and **b** is the **number a·b** given by [1]

$$\mathbf{a} \cdot \mathbf{b} = a_1 b_1 + a_2 b_2 + \ldots + a_n b_n \tag{10}$$

The dot product of two vectors is therefore a scalar (number), by definition, and not a vector. To find the dot product we multiply the corresponding components and add.

**Theorem:** if $\theta$ is the angle between the vectors **a** and **b**, then [1]

$$\mathbf{a.b} = |\mathbf{a}|\,|\mathbf{b}|\cos(\theta). \tag{11}$$

**Important property:** Two non-zero vectors **a** and **b** are orthogonal, or perpendicular, if the angle between them is $\pi/2$, therefore **a** and **b** are orthogonal if $\mathbf{a}\cdot\mathbf{b} = 0$.

### 4.4. The cross product

The cross product is the other "product" between two vectors. The caveat is that the cross product is only defined for vectors in three dimensions. An application for the cross product is torque. This is similar to work which we saw with the dot product, but the cross product results in a vector orthogonal to the two vectors that created it.

**Definition:** If $\mathbf{a} = \ < a_1, a_2, a_3 >$ and $\mathbf{b} = \ < b_1, b_2, b_3 >$ are three dimensional vectors, then the cross product of **a** and **b** is the **vector a×b** given by [1]

$$\mathbf{a} \times \mathbf{b} = \langle a_2 b_3 - a_3 b_2,\ a_3 b_1 - a_1 b_3, a_1 b_2 - a_2 b_1 \rangle \tag{12}$$

The cross product of two vectors is therefore a vector (unlike the dot product). The cross product **a×b** of two vectors **a** and **b** is a non-zero vector that is **orthogonal** to **a** and **b** (**Figure 8**).

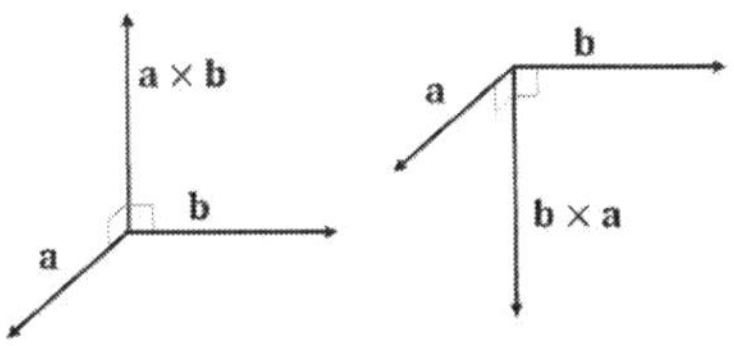


**Figure 8.** The cross product visualization.

**Theorem:** If $\theta$ is the angle between the vectors **a** and **b**, then [1]

$$|\mathbf{a} \times \mathbf{b}| = |\mathbf{a}|\,|\mathbf{b}|\sin\theta \tag{13}$$

That is, the magnitude of the resultant vector from the cross-product depends on the angle between them. The direction (while remaining orthogonal) depends on the order of the cross product, **a×b** or $\mathbf{b} \times \mathbf{a}$.

## 5. Parametric equations

An important concept within vector calculus is parameterization. In this we will create a vector function that can build a line, a curve (called a space curve) or a surface in space. The idea is that each component of the vector is a function of a single independent variable called the parameter. The parameter dictates the value of the component for each value in the domain of the parameter. The terminal point of this vector will "draw out" the line, curve or surface as the parameter varies through its domain.

$$\mathbf{r} = \langle f(t), g(t), h(t) \rangle \tag{14}$$

Example:

$$\mathbf{r} = \langle 2t + 1, 3t^2, 7 - t \rangle \tag{15}$$

### 5.1. Equation of a line

To parameterize a line we will create a vector equation in which **r** is a position vector for all points on the line. The addition of two vectors creates a vector that will "draw" the line with its terminal point based upon values of a parameter. The equation is:

$$\mathbf{r} = \mathbf{r}_0 + t\mathbf{v}[1] \tag{16}$$

The parameter is $t$, $\mathbf{v} = \ < a, b, c >$ is a vector parallel to the line, and $\mathbf{r}_0 = \ < x_0, y_0, z_0 >$ is a position vector to any single point on the line. A sketch of the mechanics of this equation is shown in **Figure 9**.

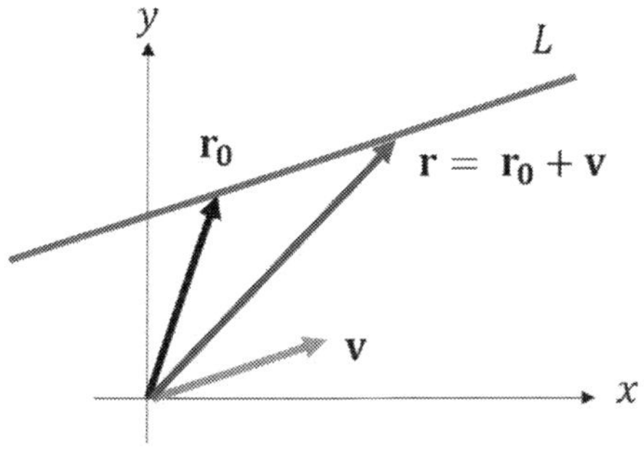


**Figure 9.** Vector equation of a line.

As $t$ varies, the terminal point of $\mathbf{r}$ "draws" the line. When we look at the resulting components from the vector equation, the parametric equations are:

$$x = x_0 + at,\ y = y_0 + bt,\ z = z_0 + ct \tag{17}$$

If we wanted to parameterize a straight line between the points (0,1) to (2,1) then we could use $x$ as the parameter, and since the $y$ value is constant across the entire line, our vector equation becomes $\mathbf{r} = \ < x, 1 >, \quad 0 \le x \le 2$.

### 5.2. Equation of a plane

The vector equation for a plane is: $\mathbf{n}\cdot(\mathbf{r} - r_0) = \ < a, b, c > \cdot < x - x_0, y - y_0, z - z_0 > \ = 0$. In this case $\mathbf{n}$ is a normal vector to the plane and $\mathbf{r}$ and $\mathbf{r}_0$ are position vectors to two points within the plane, thus their difference creates a vector in the plane. After evaluating the dot product, the scalar equation becomes:

$$a(x - x_0) + b(y - y_0) + c(z - c_0) = 0[1] \tag{18}$$

## 6. The gradient

An important calculation in vector calculus is the gradient. We will see that the gradient of a function will return a vector that points in the direction of greatest increase from any point in the function. First we will inspect where the gradient comes from.

### 6.1. Partial derivatives

As we move into three dimensions we see our functions represented as $z = f(x, y)$. Now our functions are of two variables. Remembering that a derivative is the rate of change of a function at a specific point, the partial derivative with respect to x or y becomes the rate of change at a point in either the x or y direction. There are several notations for partial derivatives [1].

$$f_x(x, y) = f_x = \frac{\partial}{\partial x} f(x, y) = \frac{\partial z}{\partial x} \tag{19}$$

$$f_y(x, y) = f_y = \frac{\partial}{\partial y} f(x, y) = \frac{\partial z}{\partial y} \tag{20}$$

The partial derivatives are found, largely, the same way as single derivatives. The only new aspect is the other variable, which we treat as a constant when taking the derivative.

Example:

- Find $\frac{\partial f}{\partial x}$, where $f(x, y) = 3x + xy^2$ $\frac{\partial f}{\partial x} = 3 + y^2$
- Find $\frac{\partial}{\partial x} f(0, 2) = 3 + 2^2 = 7$

### 6.2. Directional derivative

Now that we know how to find the rate of change in either the x or y direction, what about the other infinite directions we may be interested in? We will utilize a directional derivative to find the rate of change from a point in a particular direction.

**Theorem:** if f is a differentiable function of $x$ and $y$, then $f$ has a directional derivative in the direction of any unit vector $\mathbf{u} = \ < a, b >$ where [1]

$$D_u f(x,y) = a.f_x(x,y) + b.f_y(x,y) \tag{21}$$

The directional derivative appears to be a dot product between two vectors. In fact if we assume that it is and write the function as a dot product between two vectors the resulting equation reveals:

$$D_{\mathbf{u}} f(x,y) = \langle a,b \rangle \cdot \langle f_x(x,y), f_y(x,y) \rangle \tag{22}$$

We immediately see our vector **u**, and the second vector of partial derivatives is known as the Gradient.

### 6.3. The gradient

When talking about the gradient we will introduce an operator called the dell operator, which is similar to an upside down triangle. The del operator is a vector.

$$\nabla = \frac{\partial}{\partial x}\mathbf{i} + \frac{\partial}{\partial y}\mathbf{j} \ (5) \tag{23}$$

The gradient is the product of the del operator and the function

$$\nabla f(x,y) = \frac{\partial f}{\partial x}\mathbf{i} + \frac{\partial f}{\partial y}\mathbf{j} \tag{24}$$

Thus another way to annotate a directional derivative would be:[1]

$$D_u f(x,y)=\mathbf{u}.\nabla f(x,y) \tag{25}$$

It is important to note that the gradient of a function will point in the direction of greatest increase of a function at a specific point. The magnitude of the gradient at that point will be the greatest rate of change. When viewing a function as its level curves. The gradient will be orthogonal to the curve at any point.

Example:

Find the maximum rate of change of $f(x, y) = x^3 + 2xy^2$ at (2,1)

$$\nabla f(x,y)=\langle 3x^2+y^2,4xy\rangle,\ \nabla f(2,1)=\langle 13,8\rangle,|\nabla f(2,1)|=\sqrt{13^2+8^2}=233 \tag{26}$$

### 6.4. Tangent planes

In a similar way to finding a line tangent to a curve at a point, we can find a plane tangent to a surface at a point. The idea of a tangent plane will be applied later in the chapter.

Suppose that we have a surface $z = f(x, y)$ and we want to find the plane tangent to the surface at the point $P(x_0, y_0, z_0)$. The equation is:

$$z-z_0=f_x(x_0,y_0)(x-x_0)+f_y(x_0,y_0)(y-y_0)=\langle f_x,f_y\rangle\cdot\langle(x-x_0),(y-y_0)\rangle \tag{27}$$

If we think of the differences between the values of *x, y, z* as the change in the *x, y,* and *z* coordinates, then as that difference approaches zero, we can use the differentials for x,y,z.[1]

$$dz=f_x(x_0,y_0)dx+f_y(x_0,y_0)dy=\frac{\partial z}{\partial x}dx+\frac{\partial z}{\partial y}dy \tag{28}$$

## 7. Vector fields

We saw that a vector has both a magnitude and a direction. When we assign a vector to every point in space we build a vector field. A practical example in the real world would be wind speeds. It is easy to see that at every point the wind has both a velocity and a direction.

**Definition:** Let *D* be a set $\mathbb{R}^2$ (a plane region). A vector field on $\mathbb{R}^2$ is a function *F* that assigns to each point $(x, y)$ in D a two-dimensional vector $F(x, y)$. A vector field is expressed as follows [1].

$$F(x,y,z) = f_1(x,y,z)\mathbf{i} + f_2(x,y,z)\mathbf{j} + f_3(x,y,z)\mathbf{k} = P\mathbf{i} + Q\mathbf{j} + R\mathbf{k} \tag{29}$$

### 7.1. Curl

Perhaps, the most important attribute of a vector field is the curl. The curl will determine if there is some sort of twist or spin within the vector field. This will become very important when we start to discuss work through a vector field. The curl is determined by taking the cross product between the del operator and the vector field.

$$curl\mathbf{F} = \nabla \times \mathbf{F} \tag{30}$$

To find the actual curl vector we will find the determinate of the matrix $\nabla \times \mathbf{F}$ [2]:

$$\nabla \times \mathbf{F} = \begin{vmatrix} \hat{i} & \hat{j} & \hat{k} \\ \dfrac{\partial}{\partial x} & \dfrac{\partial}{\partial y} & \dfrac{\partial}{\partial z} \\ P & Q & R \end{vmatrix} \tag{31}$$

$$\nabla \times \mathbf{F} = \left[\frac{\partial R}{\partial y} - \frac{\partial Q}{\partial z}\right]\mathbf{i} + \left[\frac{\partial P}{\partial z} - \frac{\partial R}{\partial x}\right]\mathbf{j} + \left[\frac{\partial Q}{\partial x} - \frac{\partial P}{\partial y}\right]\mathbf{k} \tag{32}$$

If the curl results in a zero vector, then the vector field is determined to have zero curl and the vector field is conservative.

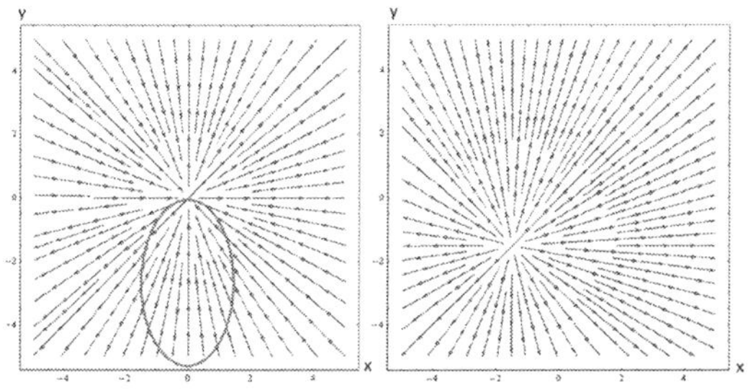


**Figure 10.** Vector fields with and without curl. Images from Wolfram Mathematica 9.0.

When we compare the plots of the two similar vector fields in **Figure 10**, we notice a subtle difference between them. In the vector field on the left there are vectors pointing towards the center as well as vectors pointing away from the center. This indicates curl. In the vector field

http://dx.doi.org/10.5772/65009

on the right, all points are pointing away from the center which indicates that there is no curl in the vector field.

### 7.2. Divergence

The divergence of a vector field can be thought of as a quantification of the rate that the vector field is expanding or contracting per unit volume at any point in the field [3]. If the divergence is equal to zero the vector field is said to be incompressible. Divergence is calculated by taking the dot product between the del operator and the vector field [2].

$$div\,\mathbf{F} = \nabla \cdot \mathbf{F} = \frac{\partial P}{\partial x} + \frac{\partial Q}{\partial y} + \frac{\partial R}{\partial z} \tag{33}$$

If the divergence at a point is positive the point is considered a source. If the divergence is negative the point is a sink. If the divergence is zero, then the point is neither a sink nor a source.

## 8. Line integrals

If we remember back to the dot product, we defined work as $W = \mathbf{F}\cdot\mathbf{d}$. We now think of a particle moving through a vector field along a curve between two points described by the vector $\mathbf{r}(t)$. The work done by the vector field on the particle as it is moving between two points can be thought of as the dot product between the vector field and the direction vector of the particle.

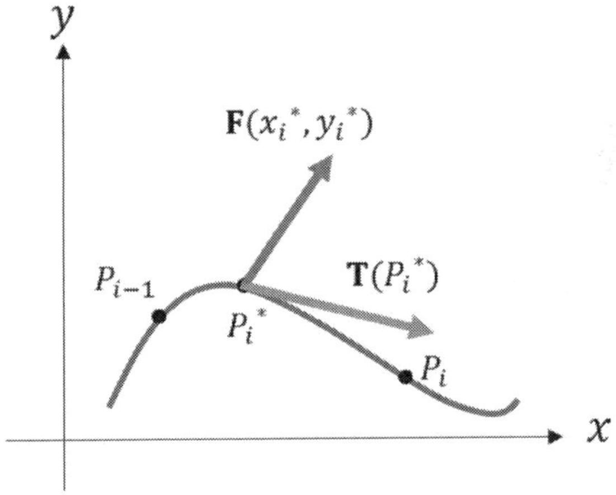


**Figure 11.** Work approximation over a line segment.

If we inspect a small segment of the curve, the work between two points can be approximated by taking the dot product of **F** and the tangent vector to the curve at a point between them, **T**(the unit tangent) or $d\mathbf{r}$ (**Figure 11**).

If $\Delta s_i$ is the length of the curve between two points, then the approximation of work over the entire curve would be the sum of work in all subsections

$$Work \approx \sum_{i=1}^{n} \mathbf{F}\left(x_i^*, y_i^*\right) \cdot \mathbf{T}\left(P_i^*\right) \Delta s_i \tag{34}$$

As $\Delta S_i$ goes to zero we have an infinite sum, or integral [4].

$$Work = \int_C \mathbf{F} \cdot \mathbf{T}\, dS = \int_C F \cdot \frac{\mathbf{r}'(t)}{|\mathbf{r}'(t)|} |\mathbf{r}'(t)|\, dt = \int_a^b \mathbf{F}(\mathbf{r}(t)) \cdot \mathbf{r}'(t)\, dt = \int_C \mathbf{F} \cdot d\mathbf{r} \tag{35}$$

### 8.1. The fundamental theorem for line integrals

We talked earlier about conservative vector fields, where the curl was zero. There is an important implication to the curl being zero; there exists a function such that the gradient of that function is equal to the vector field. This function is called the potential function.

$$\nabla \varnothing = \left\langle \frac{\partial \varnothing}{\partial x}, \frac{\partial \varnothing}{\partial y}, \frac{\partial \varnothing}{\partial z} \right\rangle = F = \langle P, Q, R \rangle \tag{36}$$

**Theorem:** Let $C$ be a smooth curve given by the vector function $r(t)$, $a \le t \le b$. Let $\emptyset$ be a differentiable function of two or three variables whose gradient vector $\nabla \emptyset$ is continuous on $C$. Then [2]

$$\int_C \nabla \varnothing \cdot d\mathbf{r} = \varnothing(\mathbf{r}(b)) - \varnothing(\mathbf{r}(a)) \tag{37}$$

Clearly, if the path is closed, where the start point is the same as the end point, the work will be zero. Another implication to the conservative vector field is path independence. This means that the work done by a conservative vector field on an object moving between two points is the same regardless of the path.

$$\int_{C_1} \mathbf{F} \cdot d\mathbf{r} = \int_{C_2} \mathbf{F} \cdot d\mathbf{r} \tag{38}$$

http://dx.doi.org/10.5772/65009

## 9. Green's Theorem

In essence, Green's Theorem allows us to calculate the work done around a closed path through a non-conservative vector field by looking at the area enclosed by the path rather than the path itself.

**Green's Theorem:** Let $C$ be a positively oriented, piecewise-smooth, simple closed curve in the $x-y$ plane and let $R$ be the region bounded by $C$. If $P$ and $Q$ have continuous partial derivatives on an open region that contains $D$, then [4]

$$\int_C \mathbf{F}\cdot d\mathbf{r} = \int_C P\,dx + Q\,dy = \iint_R \left(\frac{\partial Q}{\partial x} - \frac{\partial P}{\partial y}\right) dA \tag{39}$$

We can recognize in this integral that we are translating the line integral into a double integral of the curl in the x-y plane over the area enclosed by the curve. The positive orientation indicates that we are moving in a counter clockwise direction around the curve. Green's Theorem can also be applied to the $x-z$ plane and the $y-z$ plane if the curve or path is in either of those two planes. Green's Theorem is especially useful if the path encloses an area with a hole in it. In this case we are still integrating the curl over the area.

Example: [2]

Evaluate the given line integral around the path defined by the circles:

$$C_1 : x^2 + y^2 = 4,\ C_2 : x^2 + y^2 = 1 \tag{40}$$

$$\oint_C (4x^2 - y^3)dx + (x^3 + y^2)dy \tag{41}$$

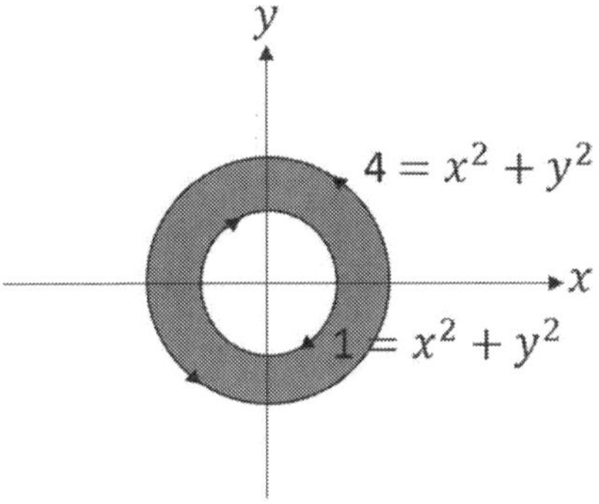

We are going to utilize the area between the two circles to solve this line integral.

By Green's Theorem:

$$\oint_C (4x^2 - y^3)dx + (x^3 + y^2)dy = \iint_R \left(\frac{\partial Q}{\partial x} - \frac{\partial P}{\partial y}\right) dA \tag{42}$$

$$\boldsymbol{F} = \left\langle \left(4x^2 - y^3\right), \left(x^3 + y^2\right) \right\rangle, \frac{\partial Q}{\partial x} = 3x^2, \frac{\partial P}{\partial y} = -3y^2 \tag{43}$$

$$\iint_R (3x^2 + 3y^2) dA \tag{44}$$

Converting to polar coordinates:

$$\int_0^{2\pi} \int_1^2 \left(3(r\cos\theta)^2 + 3(r\sin\theta)^2\right) r\, dr\, d\theta \tag{45}$$

$$\int_0^{2\pi} \int_1^2 3r^3 \ dr\, d\theta = 45\frac{\pi}{2} \tag{46}$$

## 10. Surface area

Previously we saw how to parameterize a line in which the *x*, *y* and *z* components were made to be functions of another variable (usually *t*) which created a line in space or a "space curve". Similarly we can parameterize a surface in the same way, except that we will have to utilize two variables (in this case, *u* and *v*). This will return a vector equation describing our surface as [5]:

$$\mathbf{r}(u,v) = f(u,v)\mathbf{i} + g(u,v)\mathbf{j} + h(u,v)\mathbf{k} \quad (u,v) \in D \tag{47}$$

Finding the surface area of a flat or geometric surface is relatively easy as there are known formulas for such cases. However, it is more difficult to find the surface area of a more complexly defined surface. In this case we will see that the relationship between surface area and surface integrals is similar to the relationship between arc length and line integrals.

If we have a space curve defined by the vector function $\mathbf{r}(t) = \langle x(t), y(t), z(t) \rangle$ where $a \leq t \leq b$, the length of the of the arc between $a$ and $b$ is found by: [5]

$$Arc\ Length = \int_a^b \sqrt{\left(\frac{dx}{dt}\right)^2 + \left(\frac{dy}{dt}\right)^2 + \left(\frac{dz}{dt}\right)^2}\ dt \tag{48}$$

If the curve is in the x-y plane where $y = f(x), a \leq x \leq b$ then

$$Arc\ Length = \int_a^b \sqrt{1 + \left(\frac{dy}{dx}\right)^2}\ dx \tag{49}$$

In a similar sense to the arc length, where we utilized a secant line to approximate the length of a curve, we can utilize a tangent plane to a surface to approximate the surface area.

The primary idea is that to get as close to the true surface area as possible we must use an ever increasing number of planes approximating our surface. Then as the number of planes goes towards infinity the sum of their areas approaches the area of the surface.

It is important to remember how to define a plane. Two ways that will be important to surface area are finding two vectors within the plane and by finding a point with a normal vector to the plane. In order to find two vectors within the plane we can utilize the resulting vectors from the directional derivatives of our surface at our point within the surface. With the parameterization of the surface the vectors would become $r_u$ and $r_v$. To find the normal vector to the plane containing the two directional derivative we will take their cross product.

In order to find the surface area of the plane we can scale each vector by the change of the parameter of the subinterval. We then take the cross product of the scaled vectors and by the rules for the cross product we can factor out the change in variables such that:

$$|\ \Delta u \mathbf{r}_u \times \Delta v \mathbf{r}_v| = |\mathbf{r}_u \times \mathbf{r}_v| \Delta u \Delta v \tag{50}$$

If we then sum the area of each plane, the approximation to the surface becomes:

$$\sum_{i=1}^{m} \sum_{j=1}^{n} \left| \mathbf{r}_{u_i} \times \mathbf{r}_{v_j} \right| \cdot u_i \cdot v_j \tag{51}$$

Then the surface area for our surface S is: [4]

$$A(s) = \iint_R |\mathbf{r}_u \times \mathbf{r}_v| dA = \iint_R dS \tag{52}$$

If the surface is defined in terms of $z = f(x, y)$

then the formula becomes: [4]

$$A(s)=\iint_R \sqrt{1+\left(\frac{\partial z}{\partial x}\right)^2+\left(\frac{\partial z}{\partial y}\right)^2}\, dA \tag{53}$$

## 11. Surface integrals

In this section we will see that the surface integral is very similar to what a line integral is to curves in space.

It is clear that each coordinate is a function of our two parameters. The parameterized surface, $\boldsymbol{\sigma}(u,v)$, is considered smooth if both of the partial derivatives $\boldsymbol{\sigma}_u, \boldsymbol{\sigma}_v$ are continuous and the cross product, $\sigma_u \times \boldsymbol{\sigma}_v$, is never zero on the interior of the parameter domain. This cross product is called the standard normal to $\sigma$. [6].

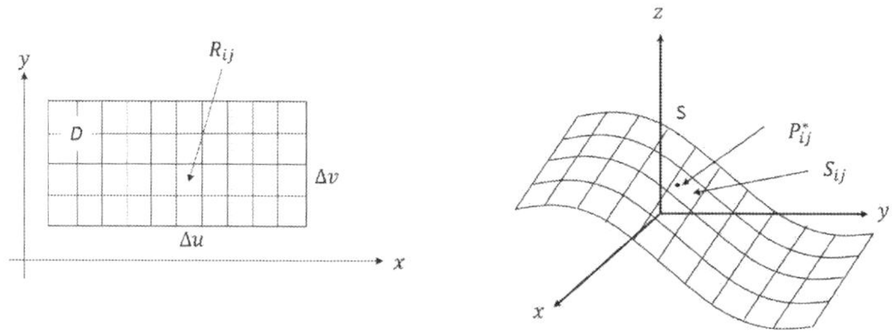


**Figure 12.** The parameter domain and corresponding surface section.

The first step is to assume the parameter domain, *D*, is an image of a function and has a rectangular shape which has been divided into rectangular sub sections identified as $R_{ij}$. The dimensions of $\Delta u, \Delta v$ and an area of $\Delta u * \Delta v$ (**Figure 12**).

Our surface is similarly divided into corresponding subsections, each defined by $S_{ij}$. If we form the Riemann sum of the product of a function $f$ evaluated at some point $P_{ij}^*$ in each subregion of our surface with the area of each subregion, $\Delta S_{ij}$. The resulting summation is: [4]

$$\sum_{i=1}^{m}\sum_{j=1}^{n} f(P_{ij}^*)\Delta S_{ij} \tag{54}$$

As the area $\Delta S_{ij}$ goes to zero and $m$ and $n$ go to infinity we have the definition of the surface integral as: [4]

$$\iint_S f(x,y,z)dS = \lim_{m,n\to\infty} \sum_{i=1}^{m} \sum_{j=1}^{n} f(P_{ij}^*)\Delta S_{ij} \tag{55}$$

Let's take a second to remember what the $dS$ was:

If we remember back to how surface area was calculated utilizing an infinite number of tangent planes, then the change in the surface area $\Delta S_{ij}$ clearly becomes $dS$ [7].

$$\iint_S f(x,y,z)dS = \iint_R f(\mathbf{r}(u,v))|\mathbf{r}_u \times \mathbf{r}_v|\, dA \tag{56}$$

Remember that $|r_u \times r_v|$ is the length of a vector normal to the surface. This allows us to compute a surface integral by converting it into a double integral over the parameter domain $D$. If, for example, our surface is defined by an equation where the parameterization can be defined by $x = x, y = y, z = h(x,y)$, then our normal vector will be: [7]

$$\mathbf{r}_x = \mathbf{i} + \frac{\partial h}{\partial x}\mathbf{k} r_y = \mathbf{j} + \frac{\partial h}{\partial y}\mathbf{k} \tag{57}$$

$$\mathbf{r}_x \times \mathbf{r}_y = -\frac{\partial h}{\partial x}\mathbf{i} - \frac{\partial h}{\partial y}\mathbf{j} + \mathbf{k} \tag{58}$$

$$\left|\mathbf{r}_x \times \mathbf{r}_y\right| = \sqrt{\left(\frac{\partial h}{\partial x}\right)^2 + \left(\frac{\partial h}{\partial y}\right)^2 + 1}\, dA \tag{59}$$

Then the resulting surface integral becomes: [7]

$$\iint_R g\left(x,y,h(x,y)\right)\sqrt{\left(\frac{\partial h}{\partial x}\right)^2 + \left(\frac{\partial h}{\partial y}\right)^2 + 1}\, dxdy \tag{60}$$

Similar formulas can be easily derived should the surface be defined in a way other than by $z = f(x,y)$.

## 12. Oriented surfaces

It is important to talk about the "orientation" of a surface. It is easy to see that there are always two vectors that are normal to any surface as they point in exactly opposite directions. We can think about a "positive" orientation being one defined by a normal vector pointing away from the origin and a "negative" orientation being defined by a normal vector pointing towards the origin. In vector calculus, the only surfaces that are used are those that are orientable. A type of surface that is not orientable, such as the Mobius Strip, is not applicable within vector calculus.

## 13. Surface integrals and flux

Suppose that we have an oriented surface, $S$, with a unit normal vector, $\mathbf{n}$ (remember, that a unit normal vector is a vector orthogonal to the surface with a length of 1 unit. If we think of this surface being within a vector field describing, perhaps, a fluid with a density function of $\rho(x, y, z)$ and a velocity field of $\mathbf{F}(x, y, z)$ flowing through our surface $S$, then the amount of the "fluid" passing through our surface at one point would be the dot product between the vector field at the point and the normal vector to the surface at that same point (**Figure 13**).

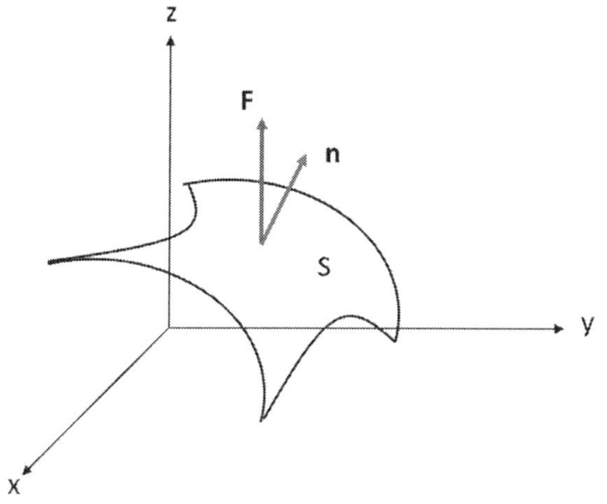


**Figure 13.** Flux visualization.

The rate of flow (mass per unit time) per unit area is $\rho\mathbf{v}$. If we divide our surface into sub regions defined by $S_{ij}$ where each subregion is small and nearly rectangular we can approximate the mass or amount of fluid crossing $S_{ij}$ in the direction of $\mathbf{n}$ per unit time by:

$$(\rho \mathbf{v} \cdot \mathbf{n}) A(S_{ij}) \tag{61}$$

If we add this mass for each sub section as the area of each subsection goes to 0 it results in an infinite sum or integral which calculates the rate of flow through $S$:

$$\iint_S (\rho \mathbf{v} \cdot \mathbf{n}) dS = \iint_S \rho(x,y,z)\mathbf{v}(x,y,z) \cdot \mathbf{n} \, dS \tag{62}$$

This type of surface integral occurs frequently in physics. If the density function is constant then we are only left with the velocity vector field.

More simply, if $\mathbf{F}$ is a continuous vector field defined on an oriented surface $S$ with a unit normal vector $\mathbf{n}$, then the surface integral of $\mathbf{F}$ over $S$ is: [4]

$$Flux = \iint_S F \cdot d\mathbf{S} = \iint_S \mathbf{F} \cdot \mathbf{n} \, dS \tag{63}$$

We are then able to translate this integral into our previously shown double integral over a region such that: [4]

$$Flux = \iint_S \mathbf{F} \cdot d\mathbf{S} = \iint_S \mathbf{F} \cdot (\mathbf{r}_u \times \mathbf{r}_v) \, dS = \iint_S (\mathbf{F} \cdot \mathbf{n}) \, dS \tag{64}$$

With regards to the actual mechanics of how this integral is calculated it is important to have an understanding of the relationship between the parts. It is also important to pay attention to how the surface is defined. Let's assume $z = f(x,y)$ is our surface. We must express the surface as a function $g(x,y,z)$. Then, $g(x,y,z) = z - f(x,y)$. The resulting normal vector of the surface becomes:

$$\mathbf{n} = \frac{\mathbf{r}_x \times \mathbf{r}_y}{|\mathbf{r}_x \times \mathbf{r}_y|} = \frac{\nabla g(x,y,z)}{|\nabla g(x,y,z)|} = \frac{-\frac{\partial g}{\partial x}\mathbf{i} - \frac{\partial g}{\partial y}\mathbf{j} + 1}{\sqrt{\left(\frac{\partial g}{\partial x}\right)^2 + \left(\frac{\partial g}{\partial y}\right)^2 + 1}} \tag{65}$$

The differential for surface area is:

$$dS = |\mathbf{r}_x \times \mathbf{r}_y| dA = |\nabla g(x,y,z)| dA = \sqrt{\left(\frac{\partial g}{\partial x}\right)^2 + \left(\frac{\partial g}{\partial y}\right)^2 + 1} \; dA \tag{66}$$

The resulting integral then becomes:

$$\iint_R \left( \mathbf{F} \cdot \frac{-\frac{\partial g}{\partial x}\mathbf{i} - \frac{\partial g}{\partial y}\mathbf{j} + 1}{\sqrt{\left(\frac{\partial g}{\partial x}\right)^2 + \left(\frac{\partial g}{\partial y}\right)^2 + 1}} \right) \sqrt{\left(\frac{\partial g}{\partial x}\right)^2 + \left(\frac{\partial g}{\partial y}\right)^2 + 1} \, dA \tag{67}$$

The denominator within the dot product and the radical within the surface differential will cancel which leaves us with the final integral: [4]

$$Flux = \iint_S \mathbf{F} \cdot d\mathbf{S} = \iint_S (\mathbf{F} \cdot \mathbf{n}) \, dS = \iint_R \mathbf{F} \cdot \left(-\frac{\partial g}{\partial x}\mathbf{i} - \frac{\partial g}{\partial y}\mathbf{j} + 1\right) dA \tag{68}$$

## 14. Stokes Theorem

Stokes' Theorem is very similar to Green's Theorem. In fact it can be thought of that Green's Theorem is a special case of Stokes' Theorem, or that Stokes' Theorem is a higher-dimensional version of Green's Theorem. What Stokes is going to do is relate a line integral with a surface integral. Stokes' Theorem will find the work done around a closed path by computing a surface integral of the curl of a vector field over a "surface cap [8]" on top of the path.

**Stokes' Theorem:** Let $S$ be a smooth oriented surface in $\mathbb{R}^3$ with a smooth closed boundary $C$ whose orientation is consistent with that of S. Assume that $F = \langle f, g, h \rangle$ is a vector field whose components have continuous first partial derivatives on $S$. Then [9]

$$\oint_C \mathbf{F} \cdot d\mathbf{r} = \iint_S curl\, \mathbf{F} \cdot d\mathbf{S} = \iint_S (curl\, \mathbf{F} \cdot \mathbf{n}) \, dS \tag{69}$$

In other words the circulation of a vector field around the boundary of an oriented surface in space in the direction counter clockwise with respect to the surface's unit normal vector equals the integral of the curl for vector field over the normal component of the surface. The surface must also be piecewise smooth.

We can think of this integral as relating a line integral around a closed path to a surface integral over a "capping surface" of the path. The interesting idea is that it doesn't matter what the surface is over the path, once the surface integral is calculated it reaches the same value!

While this result is difficult to prove in the general case, an easier intuition is gained through inspection of Green's Theorem demonstrated earlier. This can then be extended to Stokes, which is the same concept in three dimensions.

http://dx.doi.org/10.5772/65009

Example:

Evaluate $\oint_{c} \mathbf{F}\cdot d\mathbf{r}$ where $c$ is the curve of intersection of the paraboloid $z = 9 - x^2 - y^2$ and the plane $z = 5$, oriented up, through $\mathbf{F} = y\mathbf{i} + xz^3\mathbf{j} - zy^3\mathbf{k}$.

First check the curl to see if the vector field is conservative.

$curl\mathbf{F} = \left(-3zy^3 - 3xz^2\right)\mathbf{i} + \left(z^3 - 1\right)\mathbf{k}$, not conservative

We are going to utilize Stokes' Theorem to solve this problem. Remembering that we can define our surface over the path as anything, let's use the plane $z = 5$. $g(x, y, z) = z - 5, \mathbf{n} = \mathbf{k} = \frac{< 0, 0, 1 >}{\sqrt{1}}, dS = \sqrt{1}dA$

If we project the surface into the x-y plane it will form a circle. By making the substitution 5 for $z$ then $5 = 9 - x^2 - y^2$, and $x^2 + y^2 = 4$ we see our area is a circle of radius 2. We will convert $dA$ into polar coordinates.

$$\int_0^{2\pi}\int_0^{2}\left(curl\mathbf{F}\cdot\mathbf{n}\right)dS = \int_0^{2\pi}\int_0^{2}\left\langle\left(-3zy^3 - 3xz^2\right), 0, \left(z^3 - 1\right)\right\rangle\cdot\left\langle 0, 0, 1\right\rangle dA \tag{70}$$

Further substituting 5 for $z$ becomes

$$\int_0^{2\pi}\int_0^{2}(5^3 - 1)r\,dr\,d\theta \tag{71}$$

$$\int_0^{2\pi}\int_0^{2}\left(125 - 1\right)r\,dr\,d\theta = \int_0^{2\pi}\int_0^{2}\left(124\right)r\,dr\,d\theta = 124\int_0^{2\pi}\frac{2^2}{2}d\theta = 248\int_0^{2\pi}d\theta \tag{72}$$

$$Flux = 496\pi \tag{73}$$

## 15. Divergence Theorem

In this final section we move away from line integrals and again visit flux. In this case we will be looking at how a vector field moves through a solid shape in space.

The divergence theorem states: Let $E$ be a simple solid region and let $S$ be the boundary surface of $E$. Let $\mathbf{F}$ be a vector field whose component functions have continuous partial derivatives on an open region that contains $E$. Then [10]

$$Flux = \iint_S \mathbf{F} \cdot d\mathbf{S} = \iiint_E (div\ \mathbf{F})\ dv \tag{74}$$

In other words the Divergence Theorem states that the flux of $\mathbf{F}$ across the boundary surface of $E$ is equal to the triple integral of the divergence of $\mathbf{F}$ over $E$.

The Divergence Theorem is to surface integrals what Green's Theorem is to line integrals in that it allows us to convert a surface integral over a closed surface into a triple integral over a closed region.

While the equation looks to be a direct equivalence between the double and triple integrals, it is not quite that simple. If we were to try and utilize the surface integral definition for flux on a closed surface, we would have to compute the flux through each sub surface individually then add the result. In contrast, the divergence theorem allows us to find the flux through the closed surface by utilizing the divergence through the volume.

Example:

Find the outward flux of $\mathbf{F}$ through the closed surface of the cylinder $x^2 + y^2 = 4$ from $z = 0$ to $z = 3$. $\mathbf{F} = \left(6x^2 + 2xy\right)\mathbf{i} + \left(2y + x^2z\right)\mathbf{j} + \left(4x^2y^2\right)\mathbf{k}$. R is the region cut from the first octant.

$$flux = \iiint_R (div\mathbf{F})\ dV \tag{75}$$

$$div\mathbf{F} = 12x + 2y + 2 \tag{76}$$

If we project our region into the x-y plane, we see that it is a quarter circle with a radius of 2. In this problem we will utilize cylindrical coordinates to solve the integral.

$$flux = \int_0^{\frac{\pi}{2}}\int_0^2\int_0^3 (12x + 2y + 2)dzrdrd\theta = 2\int_0^{\frac{\pi}{2}}\int_0^2\int_0^3 (6x + y + 1)dzrdrd\theta$$

$$= 6\int_0^{\frac{\pi}{2}}\int_0^2 (6x + y + 1)rdrd\theta = 6\int_0^{\frac{\pi}{2}}\int_0^2 (6r^2\cos\theta + r^2\sin\theta + r)drd\theta$$

http://dx.doi.org/10.5772/65009

$$= 6\int_{0}^{\frac{\pi}{2}} 16\cos\theta + \frac{8}{3}\sin\theta + 2\,d\theta = 6\left(16 + \pi + \frac{8}{3}\right) = 112 + 6\pi \qquad (77)$$

## Author details

James Williams[1*], Britton Landry[1], Matthew Mogensen[1] and T. V. Hromadka II[1,2,3,4,5,6,7,8]

*Address all correspondence to: james.williams@usma.edu

1 Department of Mathematical Sciences United States Military Academy at West Point, NY, USA

2 Department of Civil Engineering, California State University, Fullerton, USA

3 Mathematics, Geological Sciences, Environmental Studies, California State University, USA

4 Wessex Institute of Technology, Great Britain, UK

5 American Academy of Water Resources Engineers (AAWRE), USA

6 National Council of Examiners for Engineering & Surveying (NCEES), USA

7 American Institute of Hydrology (AIH), Surface Water, Groundwater/Geo Hydrology, USA

8 National Groundwater Association, TX, USA

## References

[1] Stewart, James. *Calculus Early Transcendentals*. 8th ed. Belmont: Brooks/Cole Cengage Learning, 2015.

[2] Zill, Dennis G. Wright, Warren S. *Advanced Engineering Mathematics*. 5th ed. Burlington: Jones and Bartlett Learning, 2014.

[3] Zenor, Phillip, Edward E. Slaminka, and Donald Thaxton. *Calculus with Early Vectors*. Upper Saddle River: Prentice Hall, 1999. Print.

[4] Stewart, James. *Calculus Early Transcendentals*. 7th ed. Belmont: Brooks/Cole Cengage Learning, 2012.

[5] Hass, Joel, Maurice D. Weir, and George B. Thomas, Jr. *University Calculus Elements with Early Transcendentals*. 1st ed. Boston: Pearson Education, 2009. Print.

[6] Seeley, Robert T. *Calculus of Several Variables, an Introduction*. Glenview: Scott, Foresman, 1970. Print.

[7] Edwards, C. Henry, and David E. Penney. *Multivariable Calculus with Analytic Geometry*. 5th ed. Upper Saddle River: Prentice Hal, 1998. Print.

[8] Schey, H. M. *Div, Grad, Curl, and All That: An Informal Text on Vector Calculus*. New York: W.W. Norton & Company, 1973. Print.

[9] Briggs, William, Cochran, Lyle, Gillett, Bernard, Schulz, Eric. *Calculus for Scientists and Engineers Early Transcendentals*. P. 1147. Boston: Pearson Education, Inc. 2013.

[10] Kreyszig, Erwin. *Advanced Engineering Mathemcs*. 7th ed. New York: John Wiley & Sons, 1993. Print.

Chapter 2

# Evaluation of a Regional Climate Model for the Upper Blue Nile Region

Tadesse Terefe Zeleke, Baylie Damtie Yeshita and Fentahun Muluneh Agidew

Additional information is available at the end of the chapter

http://dx.doi.org/10.5772/64954

**Abstract**

The fourth version of the International Center for Theoretical Physics (ICTP) Regional Climate Model (RegCM4) model is evaluated for its performance over Upper Blue Nile River Basin Region (UBNRBR). The model rainfall captured the observed spatial and temporal variability of rainfall over the basin during the spring (MAM) and summer (JJA) seasons. The simulation dataset is generated using the RegCM4 for the period 1982–2009. The UBNRBR is first divided into 14 homogeneous regions using criteria including Rotated Empirical Orthogonal Function (REOF), spatial correlation and topographical features. Spatially averaged observed and simulated rainfall time series are then generated and analyzed for each region. Standardized rainfall anomalies of the observations and the simulated data are highly correlated over most of central regions, while a weak correlation is found over the east border regions of the basin. The dominant modes of rainfall variability are identified using REOF. The first leading patterns of rainfall and upper wind (averaged between 100 and 300 hpa) are highly correlated and exhibit similar features between simulated and observed dataset over the basin. Similarly, the first loading pattern of low level wind (averaged between 850 and 1000 hpa) exhibits a dipole structure across the southwestern and southeastern regions of the UBNRBR. The correlations with significant rotated principal components (RPCs) across gridded gauge, and model rainfall fields with that of low- and upper level winds show the presence of significant relationship (correlation exceeding ~0.6). Overall, that the RegCM4 shows a good performance in simulating the spatial and temporal variability of precipitation over UBNRBR.

**Keywords:** RegCMx, variability, RPC, Upper Blue Nile River Basin Region (UBNRBR)

# 1. Introduction

Regional climate models (RCMs) have become increasingly important tools to downscale global (large scale) climate information for regional applications. Numbers of studies have demonstrated the capability of regional climate models of different version (RegCMx) in downscaling global climate information for regional applications and representing details of regional climate [1–8]. Such models are driven by initial and lateral boundary conditions taken from reanalysis, observations and from global circulation model (GCM) output [3–5, 9–13]. RCMs become suitable tools for regional process studies, which increase our understanding about influence of local climatic forcing superimposed on large-scale climate variability. By coupling appropriate land surface, hydrologic or lake models with that of RCMs enables accurate simulation of detailed precipitation, temperature, surface hydrological features and other meteorological variables [14].

The sensitivity of RegCMx to dynamical configuration such as domain size, resolution and the physical parameterizations has been demonstrated in a number of studies [1, 4, 5, 11, 13]. Hence, before applying a regional climate model for regional climate variability studies, the accuracy of the model in reproducing the observed regional climate should be assessed to establish its strengths and weaknesses for the specific region [3].

Regional climate models have been utilized extensively for mid-latitude regions in wide-ranging surface climate and hydrologic process investigations. Sensitivity studies and simulation of present, past and future climate on the mesoscale and regional scale have been carried out [2, 3, 7, 11, 12, 15, 16]. Relatively few studies exist for eastern Africa climate [4, 5, 8, 17–19]. Most of the above studies of eastern Africa [except 18, 19] focused on the spring and autumn seasons, as these are the main and short rainy seasons for the equatorial Africa. The performance of the RegCM3 in reproducing the Ethiopian summer rainfall variability also evaluated [18, 19]. They found that RegCM3 not only reproduced the spatial variability of dry and wet years but also correlated well with gauge data.

Based on these considerations, in this chapter, the performance of a regional climate model (an updated version of the RegCM4) is presented [20]. Here, we briefly discuss/compare the essential atmospheric variables of observational and/with model simulation that will be necessary in the rest of the chapter to understand the various characteristics of rainfall in the basin. Such as in representing the climatology, inter/intra annual variability of atmospheric variables including rainfall and wind field with respect to relatively large set of rain gauge and satellite based observations and reanalysis datasets.

# 2. Model, data and methodology

## 2.1. Model description

The regional climate model used in this study is the ICTP RegCM4 described by Giorgi et al. [20]. It is a hydrostatic model based on the dynamical core of the Penn State/NCAR Mesoscale

Model version 4 [21] with the developments described by Giorgi et al. [11, 12]. RegCM4 includes a range of physics options, and for the present work, it uses the radiation scheme of Community Climate Model version 3 (CCM3) [22], the nonlocal planetary boundary layer scheme originally developed by Holtslag et al. [23] and later modified as described by Giorgi et al. [11, 20]. The Biosphere–Atmosphere Transfer Scheme in [24] is used for land surface process calculations. Precipitation is represented by two different terms: resolvable (large-scale non-convective) and convective (subgrid-cumulus). The resolvable scale precipitation is represented by the subgrid explicit moisture scheme [25]. For convective precipitation, three options are available: (1) the modified Anthes-Kuo scheme [12, 26], (2) the Grell scheme [21] and (3) the Emanuel scheme [27]. In addition, different schemes can be chosen for land and ocean regions [20]. After many preliminary tests, we selected the Grell scheme with the Fritsch-Chappel closure [11, 12, 28] over land and the Emanuel scheme over the ocean grid points. More information on the different physics schemes and applications of the RegCM4 model system can be found in the study of Giorgi et al. [20].

### 2.1.1. Experimental setup

The simulation/analysis period is 1982–2009, and we applied similar experimental setup with previous study; in which, its initial and lateral boundary conditions are obtained from the new ERA-Interim 1.5° × 1.5° third generation ECMWF gridded reanalysis product [29]. The sea surface temperature (SST) used to force RegCM4 is obtained from the National Oceanic and Atmospheric Administration (NOAA) weekly optimum interpolation (wk-OI) [30] on one-degree grid. The 10-min resolution global land cover characterization (GLCC) dataset for vegetation cover, land use and elevation is used as obtained from the United States Geological Survey (USGS). The model domain (**Figure 1**), the upper right panel, covers most of the African

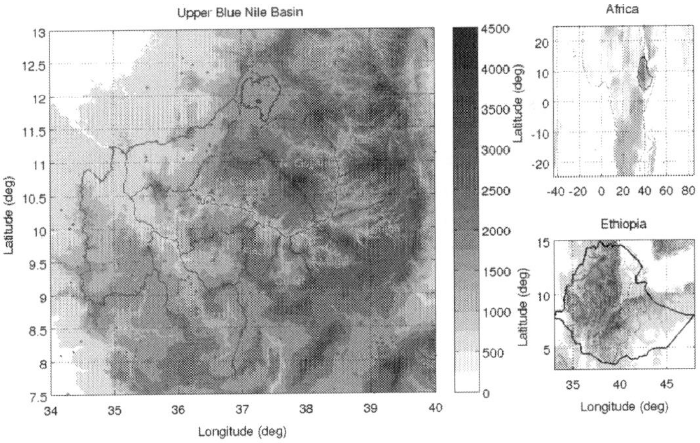


**Figure 1.** Topography (in meters) of the study area. The blue lines are the 14 homogeneous rainfall regions. The red dots represent the rainfall stations used in this study.

continent and adjacent ocean waters at a grid point spacing of 50 km [31]. This study showed that this domain size is sufficient to obtain a realistic simulation of the climate of UBNRBR.

### 2.2. Data

The station rainfall dataset used to calibrate the model output is obtained from the Ethiopian National meteorological Agency (EMA). It includes 430 unevenly distributed stations throughout the region for the period 1979–2014. The distribution of the gauges and quality control methods for the observed rainfall dataset are discussed in detail [31]. In addition to the station data described above, we use a blended gauge and satellite product: the global precipitation climatology project (GPCP) described by Adler et al. [32]. The SST is obtained from the UK Met Office Global Sea Ice and Sea Surface Temperature (HadISST2) described by Rayner et al. [33]. This product includes SST observations and satellite-derived estimates at the monthly scale with a resolution of 1° × 1°. The third generation European Centre for Medium-Range Weather Forecasts (ECMWF) 1.5° × 1.5° gridded reanalysis product of ERA-Interim [29] and National Centers for Environmental Prediction-National Center for Atmospheric Research (NCEP–NCAR) reanalysis products are also used [34].

### 2.3. Data analysis methods

To capture the patterns of co-variability of rainfall and other atmospheric variables at different stations and in the Upper Blue Nile Region, the principal component (PC) analysis (PCA) is applied to the time series. The method consists of computing the covariance matrix of the analyzed atmospheric variable dataset with the corresponding eigenvalues and eigenvectors [35]. The projection of the analyzed atmospheric fields (e.g., rainfall, wind, etc.) into the orthonormal eigenfunctions provides the PC score time series. The spatial patterns (eigenvectors), properly normalized (divided by their Euclidean norm and multiplied by the square root of the corresponding eigenvalues), are called empirical orthogonal function (EOF) or simply "loadings." The loadings in this study are the correlation values between the original data time series at each grid point and the corresponding principal component time series.

In order to extract more localized spatial patterns of variability, we apply the varimax rotation to the loadings [36–39]. Rotated empirical orthogonal function (REOF) analysis is applied to atmospheric variables such as rainfall, low-level wind (averaged between 850 and 1000 hPa), upper level wind (average of pressure levels between 100 and 300 hPa), vertical wind profile averaged over the longitude band between 35°W and 68°E and for selected oceanic basins. The region that is included in the REOF analysis of rainfall is between 34° and 40°E and 7.5° and 13°N for consistency with the gauge data.

To remove the influences of location and spread from a set of data, all atmospheric variables time series are standardized by subtracting the mean and dividing by the standard deviation. For each mode, a spatial pattern of loadings describes its area of influence and time scores that reveal the amplitude and wavelength of oscillation. Hence, we used standardized anomalies of time scores (PCs/RPCs) for correlation analysis of the dominant modes of atmospheric variables.

## 3. Result and discussion

### 3.1. Characteristics of large-scale circulation

Spring season climatological patterns of both ERA-Interim and model simulation upper level horizontal wind (**Figure 2a**, **2b** and **2e**) shows strong spatial consistency. The pattern corresponds to an anomalous southerly extension of subtropical westerly jet streams (STWJ) over northern Africa is reproduced very well by RegCM4.

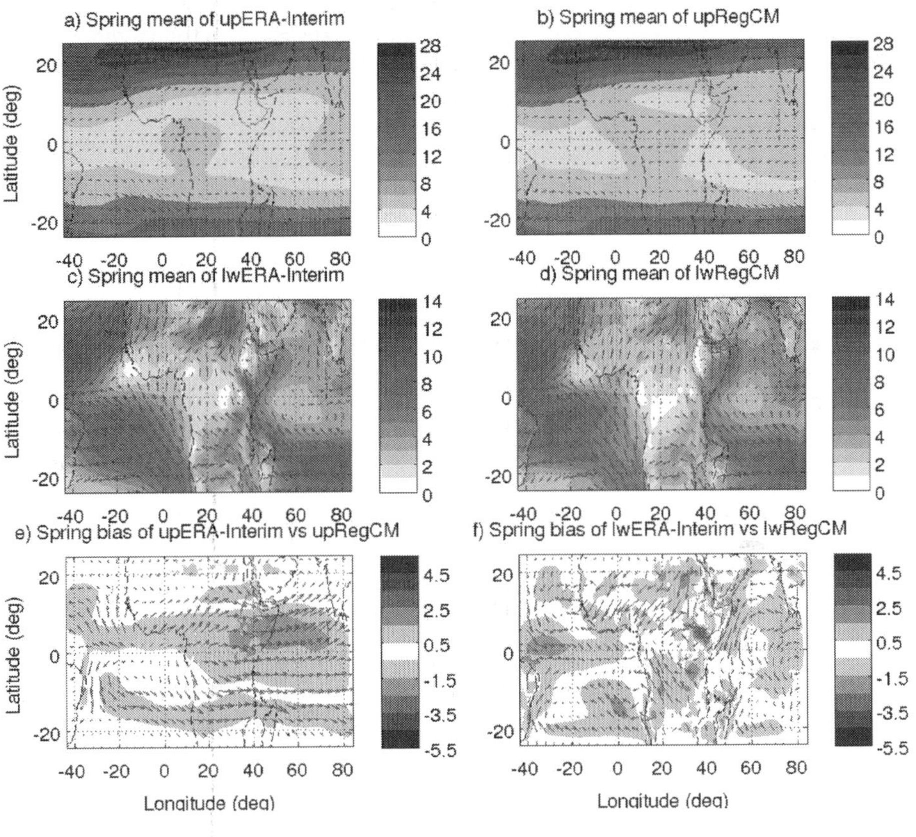


**Figure 2.** Spring (MAM) season mean horizontal wind: (a) ERA-Interim upper level (averaged between 100 and 300 hPa) wind, (b) RegCM4 upper level wind, (c) ERA-Interim low-level (averaged between 850 and 1000 hPa) wind, (d) RegCM4 low-level wind, (e) the bias of upper wind ERA-Interim vs. RegCM4 and (f) the bias of lower wind ERA-Interim vs. RegCM4.

This pattern in both ERA-Interim and model shows relatively narrow and shallow streams with maximum wind speed. The downward bent of subtropical westerly jet stream is related to a large-scale convection in the lower troposphere, which is conducive condition for spring rain getting regions of Ethiopia [19, 40]. The STWJ is formed as a result of conservation of angular momentum as the air moves from the lower latitudes to the higher latitudes [19 and reference there in]. The low-level horizontal wind climatology for the spring season in ERA-Interim and RegCM4 simulation (**Figure 2c**, **2d** and **2f**) shows a good agreement in both the magnitude and direction.

Similar, comparison of summer ERA-Interim and RegCM4 horizontal upper level winds indicates a strong similarity (**Figure 3a**, **3b** and **3e**) in representing the location and strength of the tropical easterly jet (TEJ) core, even though the jet stream is slightly stronger in the model over regions south of Chad, Central African Republic, central and eastern Ethiopia. The TEJ extends from southeast Asia across the Indian Ocean towards northeast Africa, with the jet core positioned above 10°N over the Arabian Peninsula and eastern Africa, and tilted southward over central and western Africa. The low-level wind climatology for the summer season in ERA-Interim and RegCM4 is shown in **Figure 3c**, **3d** and **3f**. Like spring, summer season also showed a good agreement in both the magnitude and direction of the east African low-level jet (EALLJ).

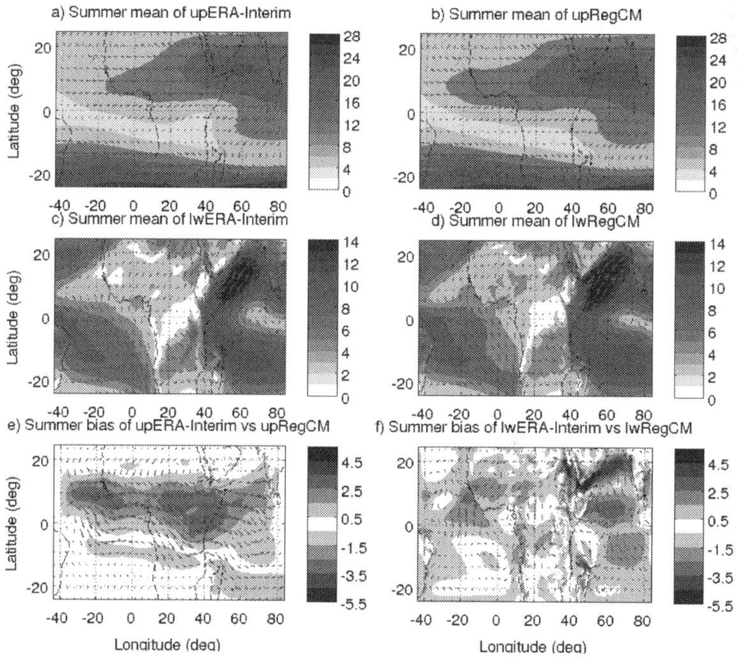


**Figure 3.** Summer (JJA) season mean horizontal wind: (a) ERA-Interim upper level wind, (b) RegCM4 upper level wind, (c) ERA-Interim low-level wind, (d) RegCM4 low-level wind, (e) the bias of upper wind ERA-Interim vs. RegCM and (f) the bias of lower wind ERA-Interim vs. RegCM.

The correlation with the time series of the wind field and the first rotated principal component (RPC1) of the upper level wind (**Figure 4a** and **4b**) reveals that the spatial pattern of the dominant mode of variability at upper levels is a dipole structure, which shows positive above ~10°N and negative below ~10°N in both model and ERA-Interim. The boundary of the dipole pattern in the model is shifted slightly northward and has stronger magnitude over southern regions.

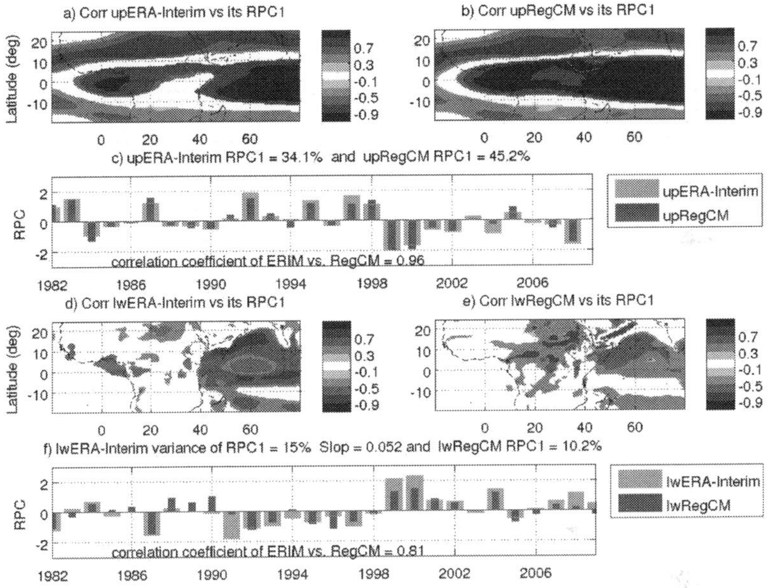


**Figure 4.** Spring (MAM) season: (a) correlation patterns of upper level horizontal ERA-Interim wind vs its dominant RPC1, (b) correlation patterns of upper level horizontal RegCM wind vs its dominant RPC1, (c) dominant time evolutions of upper level wind ERA-Interim and RegCM, (d) correlation patterns of low level horizontal ERA-Interim wind vs its dominant RPC1, (e) correlation patterns of low level horizontal RegCM wind vs its dominant RPC1 and (f) dominant time evolutions of low level wind ERA-Interim and RegCM.

The variance explained by the first RPC of the model (~45%) and ERA-Interim (~34%) are more than 1/3 of total variance. The intra-annual variability of RPC1 in ERA-Interim and RegCM4 (**Figure 4c**) shows a good agreement (correlation value of ~0.96) and the extreme years (1982/83, 1984, 1992, 1997–2000, 2008) are well captured. The similarity of variability of low-level horizontal wind patterns in **Figure 4d** and **4e** describes the performance of the model in representing the region of dominant variability in the wind field, which explains ~15% and ~10% of total variance, respectively, although small difference are observed over southwest regions of Ethiopia. Significant and high correlation (correlation value of ~0.81) of the dominant time components (RPC1s, **Figure 4f**) confirms the ability of the model to simulate the large-scale circulation. The importance/link of variability of this wind in its magnitude and direction to the Upper Blue Nile River Basin climate is discussed detail in the next Section.

Like spring, during summer season quit similar correlation (upper level wind field and its first RPC1) patterns of (**Figure 5a** and **5b**) ERA-Interim and RegCM are observed, which describes the characteristics of TEJ in both reanalysis and model simulation. The fractions of variance explained by these patterns are ~46% and 51% of ERA-Interim and RegCM, respectively. The time evolutions (RPC1s) of ERA-Interim and RegCM (**Figure 5c**) shows a good agreement (correlation value of ~0.95), and in particular, the extreme negative years (1983, 1987, 1997 and 2009) and positive years (1988, 1994 and 1998) are well captured.

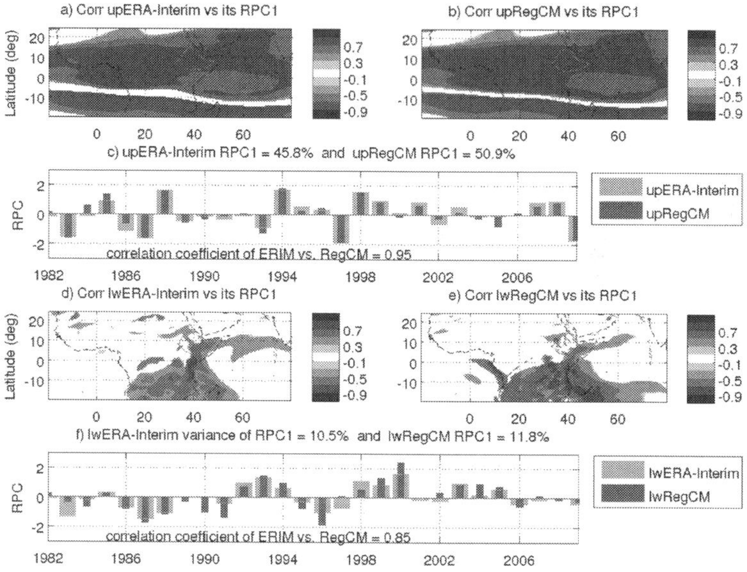


**Figure 5.** Summer (JJA) season: (a) correlation patterns of upper level horizontal ERA-Interim wind vs its dominant RPC1, (b) correlation patterns of upper level horizontal RegCM wind vs its dominant RPC1, (c) dominant time evolutions of upper level wind ERA-Interim and RegCM, (d) correlation patterns of low level horizontal ERA-Interim wind vs its dominant RPC1, (e) correlation patterns of low level horizontal RegCM wind vs its dominant RPC1 and (f) dominant time evolutions of low level wind ERA-Interim and RegCM.

The patterns of the dominant mode of variability in the lower level wind of the ERA-Interim reanalysis and RegCM4 are essentially identical (**Figure 5d** and **5e**) and show a positive loading over coast of Somalia, which corresponds to east African low-level jet (EALLJ). **Figure 5f** shows the variances explained by the first RPC of the reanalysis and models are ~11 and 12%, respectively. The correlation between the low-level ERA-Interim and RegCM4 wind is ~0.85, which shows the resemblance of the two time series (**Figure 5f**).

### 3.2. Rainfall climatology, annual cycle, and intra-annual variability

In this section, we analyze the spatial patterns, annual cycle and intra-annual variability of spring and summer rainfall over Upper Blue Nile Region. Mean seasonal rainfall over the region for the period 1982–2009 shows that the southern and central mountainous regions receive on average more than 12 mm/day during summer and small (1–2 mm/day) amount of rainfall during spring seasons (**Figure 6**). The western and eastern regions, which are semiarid, receive comparably less precipitation during these seasons. The model reproduces reasonably well this climatological pattern of rainfall, although with positive and negative biases over the western mountainous regions and some isolated areas of Upper Blue Nile River Basin region. It exhibits also a central-east/west gradient where rainfall decreases from ~12 to less than ~7 mm/day. RegCM4 forced by ERA-Interim reanalysis capture the location of higher precipitation rates in the southwest, central and northeastern region better than GPCP and CRU dataset. We also note that the GPCP dataset show relatively low precipitation amounts over the southern and central mountain regions of the basin with respect to gauge.

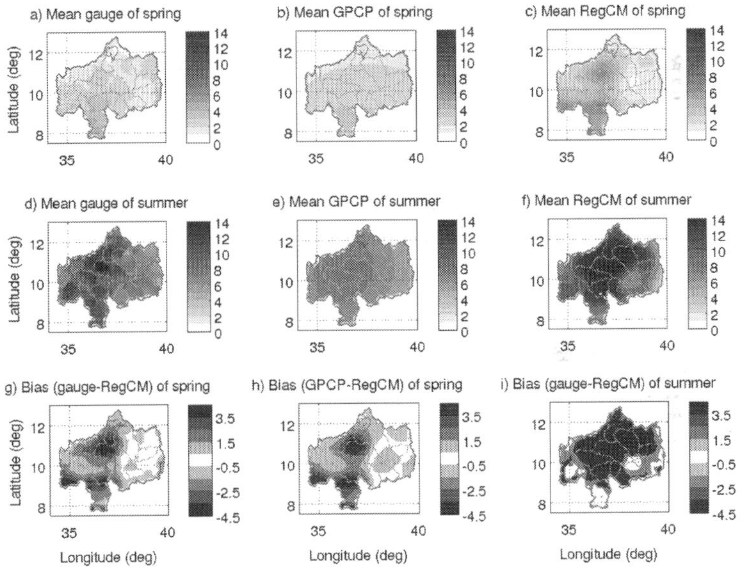


**Figure 6.** Rainfall: (a) spring mean gauge rainfall, (b) spring GPCP mean rainfall, (c) spring RegCM mean rainfall, (d) summer mean gauge rainfall, (e) summer GPCP mean rainfall, (f) summer RegCM mean rainfall, (g) bias of gauge vs. RegCM during spring, (h) bias of GPCP vs RegCM during spring and (i) bias of gauge vs. RegCM during summer.

**Figure 7** shows the mean annual cycle for the homogeneous rainfall regions of the basin using different sets of observations (gauge, GPCP and CRU) and RegCM4 simulation. The annual cycle values are averaged for each homogeneous region of the Upper Blue Nile River Basin over the whole observation and simulation periods. Over Tana subregions (**Figure 7a**), which lie over the "Semien Mountain," the model captures the summer monsoon rainfall and the pre- (May) and post-monsoon rain, although some slight differences in the intensities among the observed and model estimates are observed.

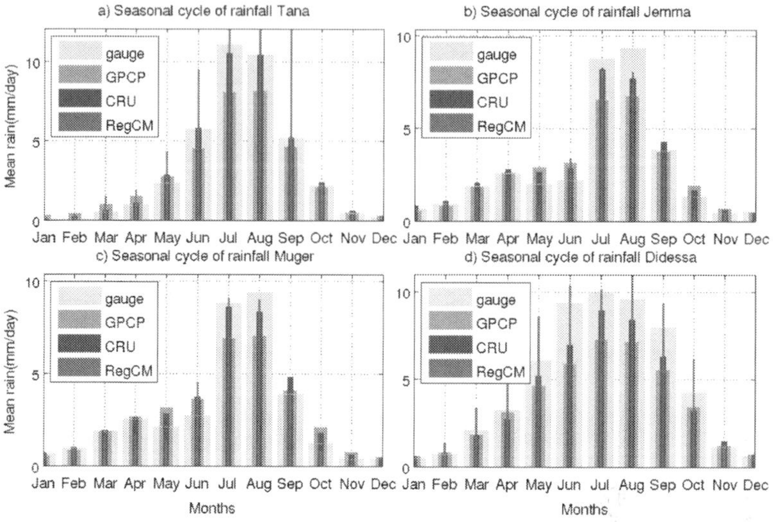


**Figure 7.** Areal averaged of homogeneous regions seasonal rainfall or annual cycle.

For Jemma and Muger regions (**Figure 7b** and **7c**), gauge, GPCP and CRU observations exhibit a maximum July–August and small rain during spring season. The model captures this cycle except small underestimation of the rainfall amount (compared to observational dataset or gauge). The Didessa region of the basin (**Figure 7d**) shows unimodal, but longer rainy season almost throughout the year maximum between May and September. However, there is a wide spread in the magnitude and phase of the precipitation maxima across these datasets, with the gauge showing the largest magnitudes, GPCP the smallest and CRU some intermediate values. RegCM4 also captures well the long seasonality, and slightly overestimate precipitation amounts throughout the year. Overall, RegCM4 performs well in reproducing the seasonal cycle of precipitation in all regions, except for an overestimation over the southwestern and central regions, where the rainfalls in almost all months is systematically underestimate over the eastern regions. We note that GPCP had better agreement in the magnitude, and especially, the phase of the rainy season peaks and corresponding breaks than CRU with respect to gauge dataset. Most previous studies of RegCM3 using a smaller domain centered over eastern Africa found difficulties to correctly reproduce the precipitation patterns. For example, in [18] performed 18 years of simulation with RegCM3 over eastern Africa and reported overestimation by >26% precipitation in Ethiopia, using the Grell/Emanuel convective scheme. In reference [8] indicated some deficiencies in capturing short east Africa rainy season of the observed rainfall over the Kenya Highlands and Lake Victoria Basin using RegCM3. Sun et al. [4] showed also some deficiencies over the Congo–Angola Basin and Kenya Highlands and the monsoon flow during the same period was stronger than observed.

**Figure 8** shows the spatial patterns of correlation coefficient between the first dominant summer RPC1 of observational rainfall and the corresponding raw summer mean rainfall time series at each grid point over the basin region. According to the result, the northeastern regions have high coefficient of variation in rainfall during summer season. The variance explained over this region is ~15, ~26, ~13 and ~33% of the total variance using gauges, GPCP, RegCM4 and CRU, respectively. The pattern of gauge (**Figure 8a**) is narrower when we compare with respect to GPCP and CRU and similar with model simulation (**Figure 8a**, **8b**, **8c** and **8d**). The model and gauge shows significant negative correlation over small western regions of the basin, unlike gauge and CRU. The corresponding RPC1 of RegCM4 significantly correlated with RPC1 of gauge (correlation between RPC1s of gauge and RegCM4 is ~0.67). Similarly, RegCM4-RPC1 correlated significantly with GPCP-RPC1 (correlation between them is ~0.82) and the two observational RPCs (gauge and GPCP) correlated with a magnitude of ~0.88. The dominant RPC of RegCM4 captured correctly the extreme positive years (e.g. 1983, 1989 and 1990) and extreme negative years (e.g., 1984, 2000 and 2008/2009) with that of corresponding observational RPCs.

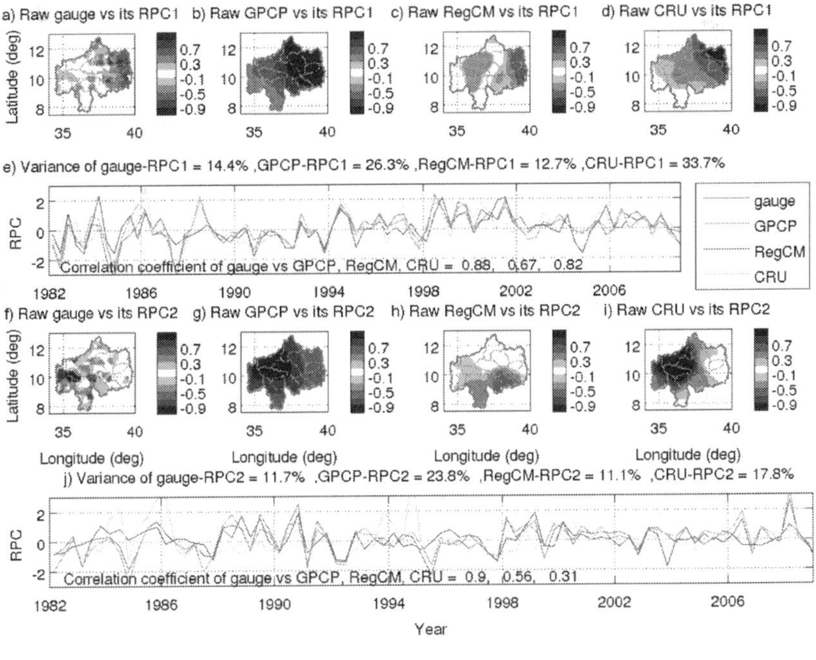


**Figure 8.** Summer season, correlation patterns of rainfall first dominant RPC1 with raw rainfall: (a) gauge, (b) GPCP, (c) RegCM, (d) CRU, (e) the time series of the RPC1 for the observations and the RCM, (f–i) correlation patterns of rainfall RPC2 with raw rainfall, of gauge, GPCP, RegCM and CRU, respectively and (j) the time series of the RPC2 for the observations and the RCM.

The second patterns of correlation between second RPC2 of rainfall and summer mean rainfall by gauge, GPCP and RegCM4 are shown in **Figure 8f**, **8g** and **8h**, respectively. The patterns in all dataset indicate a strong correlation over northwestern and western regions of Upper Blue Nile River Basin. About 12, ~24 and ~11% variances are explained by the patterns out of total variance over southwestern region using gauge, GPCP and RegCM4, respectively. The variance explained by RegCM4 over this region is smaller than GPCP, but relatively same compared to gauge. RegCM4 and gauge RPC2 negatively correlated with their corresponding rainfall over northeastern region unlike GPCP. The second gauge RPC is significantly correlated with GPCP and RegCM4, with a correlation with ~0.9 and ~0.56, respectively. Similar to the first RPC, the observational RPC2 extreme positive years (e.g., 1983, 1988, 1996 and 2007) and extreme negative years (e.g., 1996, 2002, and 2008/9) are clearly reproduced by the model. Our simulation showed relatively good performance when we apply dry Grell over land and Emanuel over Ocean and we used ERA-Interim at the lateral boundaries of the simulation.

Observed areal averaged standardized anomalies of each homogeneous spring mean rainfall (not shown here) time series are significantly (a significance level of 95%) correlated with the corresponding simulation dataset. The result indicates the model standardized precipitation anomaly is highly correlated with the corresponding gauge and GPCP datasets. On the contrary, weak correlation over most of homogeneous regions with CRU dataset, which may be because of high spatial difference over the region, using small number of stations may bring such result between simulation and CRU dataset unlike GPCP and gauge. Similar results were reported by Tsidu [41].

Overall the RegCM4 simulates fairly the observed multi-scale spatial and inter-/intra-annual temporal variability of climate in UBNRBR (correlation with gauge >0.7). We also noticed GPCP represents the observed multi-scale variability better (correlation with gauge >0.83) than CRU for both homogeneous areal mean standardized time series and dominant RPCs.

## 4. Summary and conclusion

In this study, we investigated the ability of the RegCM4 to simulate the multi-scale spatial and temporal variability of large-scale circulation and rainfall for spring and summer season. The main large-scale circulation that connected with the generation of rainfall during summer season over the basin (such as TEJ and EALLJ) is realistically simulated. Comparison of ERA-Interim and RegCM4 horizontal upper level winds indicates a strong similarity in representing the location and strength of the TEJ core, even though the jet stream is slightly stronger in the model over regions of south Chad, Central African Republic, central and eastern Ethiopia. Model simulated low-level horizontal wind has a good agreement during summer season with reanalysis wind dataset in both the magnitude and direction over Ethiopia in general and the basin in particular. The correlation with simulated and ERA-Interim first dominant rotated principal components (~0.95) of upper level horizontal winds of summer is significant and high in magnitude. We also notice that similar patterns of RPC1s show the ability of the model to capture the features of this wind, which is highly connected with the rainfall variability over most of Upper Blue Nile basin regions during summer season. The low-level horizontal wind spatial patterns of dominant variability and high magnitude (correlation value of ~0.86) of the corresponding RPCs also confirm the performance of the model to capture the main features of rain generating mechanisms over the basin.

Upper Blue Nile River Basin summer climate variability from different observation datasets along with the performance of the regional climate model (RegCM4) in reproducing this variability is also assessed. The observed rainfall datasets have indicated that central mountainous regions, S. Gojjam, Beles, Wonbera, Anger, southern of Dabus, Didessa and Tana basin receive on average more than 12 mm/day of rainfall during the summer season, on the contrary Beshilo, Welaka, and Jemma regions, which are semiarid, receive comparably less precipitation during this season. Similar climatological pattern of rainfall is shown using GPCP, although with positive and negative biases over the western mountainous regions and some isolated lowland areas, respectively.

The mean annual cycle for the homogeneous rainfall regions using different sets of observations (gauge, GPCP and CRU) and RegCM4 simulation shows RegCM4 performs well in reproducing the seasonal cycle of precipitation over all regions, except for an overestimation over the southwestern regions, where the rainfalls in almost all months and systematically underestimate the eastern regions. We have noted that GPCP had better agreement in magnitude, and especially, the phase of the rainy season peaks and corresponding breaks than CRU with respect to gauge dataset.

The correlation coefficients between simulated and observed rainfall anomalies normalized by the standard deviation over the 14 climate subregions during spring and summer seasons and between the first two dominant RPCs show the ability of RegCM4 simulation to reproduce intra-annual variability of rainfall over subregions of the basin. The first dominant pattern of observational dataset which explains the east and western regions for spring and summer seasons, respectively, is captured correctly by RegCM4 simulation with corresponding RPCs significant correlation (correlation >0.6). Similarly, the second dominant variability regions (spring-eastern and summer-western) are simulated fairly with significant correlation with corresponding RPCs (correlation >0.56) including extreme years.

The simulated climatologies and intra-annual variability of different homogeneous climate subregions of the basin are consistent with the observed variables in representing these subregions. In particular, the model reasonably reproduces the observed rainfall and wind field climatology and intra-annual variability during both seasons. Conversely, the model has evidently weak representation of variability of temperature during both seasons with respect to station and CRU temperatures and better with respect to ERA-Interim. The spatial and temporal characteristics of climate in the region of the Upper Blue Nile Basin have been presented. Rainfall is highly seasonal, roughly highest percent of annual rainfall occurring between June and September.

The model captures the general patterns of the observed rainfall distributions, in particular the ICTZ position and intensities, although it is overestimated by the model as compared to the observation datasets. Both gauge and GPCP show the highest correlations with regard to the two dominant RPCs as compared to the rest of the datasets, but the pattern of variability of the model is best agreed with the gauge intra-annual variability in both summer and spring seasons. The RegCM4, compared to the observations, shows a little more bias in rainfall estimation than temperature. This shows that temperature variability depends more on local process, hence RegCM4 correct the temperature and make it to have better representation of observed variability.

## Acknowledgements

The author like to acknowledge greatly National Meteorological Services of Ethiopia. The authors would like to thank NOAA/OAR/ESRL for providing the GPCP data via Web site at http://www.esrl.noaa.gov/psd/, ECMWF for providing the ERA-Interim dataset. The SST data are available from the National Ocean and Atmosphere Administration website at http://www.cdc.noaa.gov. The author would also like to extend his gratitude to Blue Nile Water Institute (BNWI) and David Tadesse for financial support to complete this work.

## Author details

Tadesse Terefe Zeleke[1*], Baylie Damtie Yeshita[2] and Fentahun Muluneh Agidew[2]

*Address all correspondence to: tadessekid@yahoo.com

1 Blue Nile Water Institute, Bahir Dar University, Bahir Dar, Ethiopia

2 Washera Geospace and Radar Science Laboratory, Bahir Dar University, Bahir Dar, Ethiopia

## References

[1] Giorgi F, Mearns L. Introduction to special section. Regional climate modeling revisited. J Geophys Res. 1999; 104:6335–6352.

[2] Giorgi F, Shields C. Tests of precipitation parameterizations available in latest version of NCAR regional climate model (RegCM) over continental United States. J Geophys Res. 1999; 104(D6):6353–6375.

[3] Small EE, Giorgi F, Sloan LC. Regional climate model simulation of precipitation in central Asia: mean and interannual variability. J Geophys Res. 1999; 104:6563–6582.

[4] Sun L, Semazzi F, Giorgi F, Ogallo LA. Application of the NCAR regional climate model to eastern Africa. Part I: simulation of the short rains of 1988. J Geophys Res. 1999a; 104:6529-6548.

[5] Sun L, Semazzi F, Giorgi F, Ogallo LA. Application of the NCAR regional climate model to eastern Africa. Part II: simulation of interannual variability of short rains. J Geophys Res. 1999b;104:6549–6562.

[6] Wang Y, Leung LR, McGregor JL, Lee DK, Wang WC, Ding Y, Kimura F. Regional climate modeling: progress, challenges and prospects. J Meteorol Soc Jpn. 2004; 82:1599–1628.

[7] Pal JS, Giorgi F, Bi X, Elguindi N, Solomon F, Gao X, Francisco R, Zakey A, Winter J, Ashfaq M, Syed F, Bell JL, Diffanbaugh NS, Kamacharya J, Konare A, Martinez D, da Rocha RP, Sloan LC, Steiner A. The ICTP RegCM3 and RegCNET:regional climate modeling for the developing world. Bull Am Meteorol Soc. 2007; 88:1395–1409.

[8] Anyah RO, Semazzi F. Variability of East African rainfall based on multiyear RegCM3 simulations. Inter J Climatol. 2007; 27:357–371.

[9] Dickinson RE, Errico RM, Giorgi F, Bates GT. A regional climate model for the western United States. Clim Change. 1989; 15:383-422.

[10] Giorgi F, Bates GT. The climatological skill of a regional model over complex terrain. Mon Wea Rev 1989; 117:2325–2347.

[11] Giorgi F, Bates GT, Nieman SJ. The multi-year surface climatology of a regional atmospheric model over the western United States. J Clim. 1993a; 6:75–95.

[12] Giorgi F, Marinucci MR, Bates GT. Development of a second generation regional climate model (RegCM2). I: boundary layer and radiative transfer processes. Mon Weather Rev. 1993b; 121:2794–2813.

[13] Denis B, Laprise R, C^ot^e J. Donwscaling ability of one-way nested regional climate models: the big-brother experiment. Climate Dynamics. 2002; 18:627–646.

[14] Leung LR, Ghan SJ. Pacific Northwest climate sensitivity simulated by a regional climate model driven by a GCM. Part I: control simulation. J Clim. 1999; 12:2010–2030.

[15] Giorgi F. Simulation of regional climate using a limited area model nested in a general circulation model. J Clim. 1990; 3:941–963.

[16] Giorgi F, Bi X. A study of internal variability of a regional, climate model. J Geophys Res. 2000; 105(D24):29503–29521.

[17] Davis N, Bowden J, Semazzi F, Xie L, Onol B. Customization of RegCM3 regional climate model for eastern Africa and a tropical Indian Ocean domain. J Clim. 2009; 22:3595–3616. doi:10.1175/2009JCLI2388-1.

[18] Segele ZT, Leslie LM, Lamb PJ. Evaluation and adaptation of a regional climate model for the Horn of Africa: rainfall climatology and interannual variability. Inter J Climatol. 2008; 29(1), 47–65 doi:10.1002/joc.1681.

[19] Zeleke T, Giorgi F, Mengistu Tidu G, Diro GT. Spatial and temporal variability of summer rainfall over Ethiopia from observations and a regional climate model experiments. Theor Appl Climatol. 2013; 111:665–681. doi 10.1007/s00704-012-0700-4.

[20] Giorgi F, Coppola E, Solmon F, Mariotti L, Sylla MB, Bi X, Elguindi N, Diro GT, Nair V, Giuliani G, Turuncoglu UU, Cozzini S, Guettler I, O'Brien TA, Tawfik AB, Shalaby A, Zakey AS, Steiner AL, Stordal F, Sloan LC, Brankovic C. RegCM4: model description and preliminary tests over multiple CORDEX domains. Clim Res. 2012; 52:7–29.

[21] Grell GA, Dudhia J, Stauffer DR. Description of the fifth generation Penn State/NCAR mesoscale model (MM5). Technical note. 1994; 121p. NCAR/TN-398-STR.

[22] Kiehl JT, Hack JJ, Bonan GB, Boville BA, Briegleb BP, Williamson DL, Rasch PJ. Description of the NCAR community climate model (CCM3). Technical Report. 1996;152p. TN-420-STR. NCAR, Boulder.

[23] Holtslag AAM, DeBruin EIF, Pan HL. A high resolution air mass transformation model for short-range weather forecasting. Mon Weather Rev. 1990; 118:1561–1575.

[24] Dickinson RE, Henderson SA, Kennedy PJ. Biosphere-Atmosphere Transfer Scheme (BATS) version 1E as coupled to the NCAR Community Climate Model. Technical Note. 1993;72p. NCAR/TN-387 + STR.

[25] Pal JS, Small EE, Eltahir EAB. Simulation of regional-scale water and energy budgets: representation of sub-grid cloud and precipitation processes within RegCM. J Geophys Res. 2000; 105:29579–29594.

[26] Anthes RA. A cumulus parameterization scheme utilizing a one dimensional cloud model. Mon Weather Rev. 1977; 105:270–286.

[27] Emanuel KA, Rothman MZ. Development and evaluation of a convection scheme for use in climate models. J Atmos Sci. 1999; 56:1756–1782.

[28] Fritsch JM, Chappell CF. Numerical prediction of convectively driven mesoscale pressure systems. Part I: convective parameterization. J Atmos Sci. 1980; 37:1722–1733.

[29] Dee DP, Uppala SM, Simmons AJ, Berrisford P, Poli P, Kobayashi S, Andrae U, Balmaseda MA, Balsamo G, Bauer P, Bechtold P, Beljaars A, van de Berg L, Bidlot J, Bormann N, Delsol C, Dragani R, Fuentes M, Geer AJ, Haimberger L, Healy SB, Hersbach H, Hólm EV, Isaksen L, Kållberg P, Köhler M, Matricardi M, McNally AP, Monge-Sanz BM, Morcrette JJ, Park BK, Peubey C, de Rosnay P, Tavolato C, Thépaut JN, Vitart F. The ERA-Interim reanalysis: configuration and performance of the data assimilation system. Q J R Meteorol Soc. 2011; 137:553–597. doi:10.1002/qj.828.

[30] Reynolds RW, Rayner NA, Smith TM, Stokes DC, Wang W. An improved in situ and satellite SST analysis for climate. J Clim. 2002; 15:1609–1625.

[31] Zeleke T, Giorgi F, Mengistu Tidu G, Diro GT. Spatial and temporal variability of summer rainfall over Ethiopia from observations and a regional climate model experiments. Theor Appl Climatol. 2013. doi:10.1007/s00704-012-0700-4.

[32] Adler RF, Huffman GJ, Chang A, Ferraro R, Xie P, Janowiak J, Rudolf B, Schneider U, Curtis S, Bolvin D, Gruber A, Susskind J, Arkin P. The Version 2 Global Precipitation Climatology Project (GPCP) monthly precipitation analysis (1979–present). J Hydrometeorol. 2003; 4:1147–1167.

[33] Rayner NA, Brohan P, Parker DE, Folland CK, Kennedy JJ, Vanicek M, Ansell TJ, Tett SFB: Improved analyses of changes and uncertainties in sea surface temperature measured in situ since the mid-nineteenth century: The HadSST2 Dataset. J Clim. 2006; 19:446–469. doi:10.1175/JCLI3637.1.

[34] Kistler R, Kalnay E, Collins W, Saha S, White G, Woollen J, Chelliah M, Ebisuzaki W, Kanamitsu M, Kousky V, Dool H, Jenne R, Fiorino M. The NCEP-NCAR 50-year reanalysis: monthly means CD-ROM and documentation. Bull Am Meteor Soc. 2001; 82:247–268.

[35] Navarra A, Simoncini V. A guide to empirical orthogonal functions for climate data analysis. Springer: Dordrecht; 2010. doi:10.1007/978-90-481-3702-21.

[36] Richman MB. Rotation of principal components. J Climatol. 1986; 6:293–335.

[37] Jolliffe IT. Rotation of principal components: some comments. J Climatol. 1987; 7:507–510. doi:10.1002/joc.3370070506.

[38] Rencher AC. Multivariate statistical inference and applications. Wiley: New York; 1998.

[39] Von Storch H, Zwiers FW. Statistical analysis in climate research. Cambridge University Press. 1999; 484 p. ISBN 0521 450713.

[40] Camberlin P, Philippon N. The East African March-May Rainy season: Associated Atmospheric dynamics and predictability over the 1968-97 periods. J Clim. 2002; 15:1002–1019.

[41] Tsidu M. High-resolution monthly rainfall database for Ethiopia: homogenization, reconstruction, and gridding. J Clim. 2012; 25:8422–8443.

Chapter 3

# Energy-Water Balance and Ecosystem Response to Climate Change in Southwest China

Jianhua Wang, Yaohuan Hang, Dong Jiang and Xiaoyang Song

Additional information is available at the end of the chapter

http://dx.doi.org/10.5772/64960

**Abstract**

It is important to highlight energy-water balance and ecosystem response to climate changes. The change of water-energy balance and ecosystem due to climate change will affect the regional ecological and human living significantly, especially in Southwest China which is an ecologically fragile area. This chapter presents the retrieval methodology of parameters (reconstruction of vegetation index, land cover semi-automatic classification, a time series reconstruction of land surface temperature based on Kalman filter and precipitation interpolation based on thin plate smoothing splines), time-series analysis methodology (land cover change, vegetation succession and drought index) and correlate analysis methodology (correlation coefficient and principal component analysis). Then, based on the above method, remote sensing data were integrated, a time series analysis on a 30-year data was used to illustrate the water-energy balance and ecosystem variability in Southwest China. The result showed that energy-water balance and ecosystem (ecosystem structures, vegetation and droughts) have severe response to climate change.

**Keywords:** energy-water balance, climate change, ecosystem, droughts estimation, vegetation index, time series analysis, land cover change

## 1. Methodology

Southwest China consists of the municipality of Chongqing and the four provinces of Guangxi, Yunnan, Guizhou and Sichuan, 21°07′N to 34°14′N, 97°30′E to 112°05′E (**Figure 1**). Southwest China, which was considered to be rich in water resources, has suffered climate changes in the last decades. It is particularly important to highlight energy-water balance and ecosystem response to climate changes in Southwest China using the remote sensing techniques and Geographic Information System(GIS). This chapter introduced the methodologies for retrieving parameters of land cover, vegetation index, gridded precipitation, land surface temperature, which can be used as the factors of climate change and ecosystem. Furthermore, the climate change response of droughts estimation methods was presented. The time series analysis of water-energy balance and ecosystem variability of Southwest China in recent 30 years showed that ecosystem structures, vegetation and droughts were affected by the change of climate. The following paragraphs describe the recent achievements in more detail.

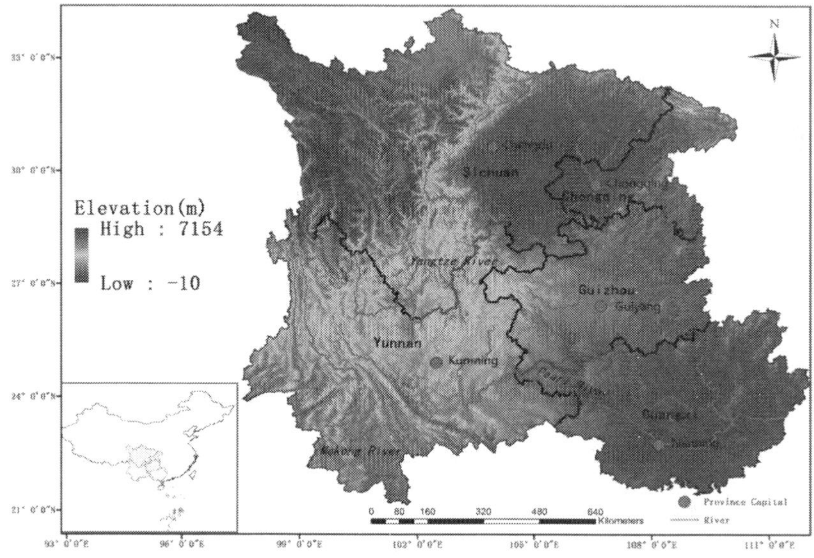


**Figure 1.** Location of Southwest China.

## 1.1. Retrieval of parameters of energy-water balance and ecosystem

### *1.1.1. Ecosystem (vegetation index): reconstruction of vegetation index based on Savitzky-Golay filter*

Terrestrial ecosystem is extremely sensitive to climate change, especially the corresponding change of the surface vegetation is the most significant. Vegetation is the important feature of the land surface and is the core and function parts of the biosphere and ecological system. It is widely accepted that energy and water together drive diversity and form of vegetation [1, 2]. In order to show the vegetation information, many kinds of vegetation index (VI), such as Normalized Difference Vegetation Index(NDVI), Soil-Adjusted Vegetation Index(SAVI), Transformed Soil-Adjusted Vegetation Index(TSAVI), Modified Soil-Adjusted Vegetation

Index(MSVAI), Difference Vegetation Index(DVI), Green Vegetation Index (GVI), Perpendicular Vegetation Index(PVI), Enhanced Vegetation Index(EVI), etc., have been developed [3]. And most of these attempts focused on the differences of the absorptive and reflective electromagnetic spectrum properties between the visible (VIS) and near-infrared (NIR) portions. According to some researchers, the time series Vis can derived from NOAA/AVHRR, SPOT/VEGETATION, TERRA or AQUA/MODIS, and the time series Vis can detect long-term development of vegetation, evapotranspiration, drought, plants phenology, corn yield, land-use/cover changes, and terrestrial ecosystems at different spatial and temporal resolutions globally [4–7]. Theoretically, because of the subtle vegetation canopy changes with respect to time, a generalized VI temporal profile is continuous and smooth [8]. However, the time series VI data always fluctuate with remarkable rises and falls which is the result of disturbances possibly caused by cloud contamination, atmospheric variability and bi-directional effects. Besides, there are a lot of no-data pixels in some VI data products because of hardware or human factors [9]. For instance, the VI datasets of MOD13A2 of MODIS need cooperate with its quality assurance (QA) data layer for application. The reliability and quality of VI values can be checked from the QA data layer. A series of approaches have been developed to reduce the errors of the noises in VI data and to reconstruct high quality time series datasets. Among them, a simple algorithm based on Savitzky-Golay filter explored by Chen was thought to be most efficient [10].

To reconstruct the high-quality MODIS EVI time series data, the Savitzky-Golay filter was used based on two hypotheses proposed by Chen: (1) The EVI data from a satellite sensor are primarily related to vegetation changes. Such as, an EVI time series follows annual cycle of vegetation growth and decline; (2) Clouds and poor atmospheric conditions made an EVI time series incompatible with the gradual process of vegetation change, because these usually depress EVI values and cause sudden drops in EVI which are regarded as noise and removed [10].

The Savitzky-Golay filter process requires continuous data, while the MOD13A2-EVI is not continuous as it contains poor quality pixels (for example, cloudy, not being processed). Based on the quality assurance (QA) dataset generated in preprocessing of MODIS EVI data in spatial, we produce a continuous dataset by interpolating the poor quality pixels of EVI data. Firstly, we searched eight pixels with good QA nearest a given pixel with poor quality and recorded the value of those eight pixels and their distance to the given poor quality pixel. **Figure 2** describes the eight pixels by eight different directions searching.

After that, the inverse distance weighted interpolation method (IDW) is applied to compute the values of the given pixel. According to the definition of IDW, we can define EVI as:

$$\lambda_i = \frac{d_i^{-1}}{\sum_1^8 d_i^{-1}} \tag{1}$$

$$EVI = \sum_{i=1}^{8} \lambda_i EVI_i \tag{2}$$

where *EVI* is EVI value of poor quality pixel after interpolation and *EVIi* is the EVI value of a good pixel. The continuous initial EVI time series is recorded as *EVI0.*

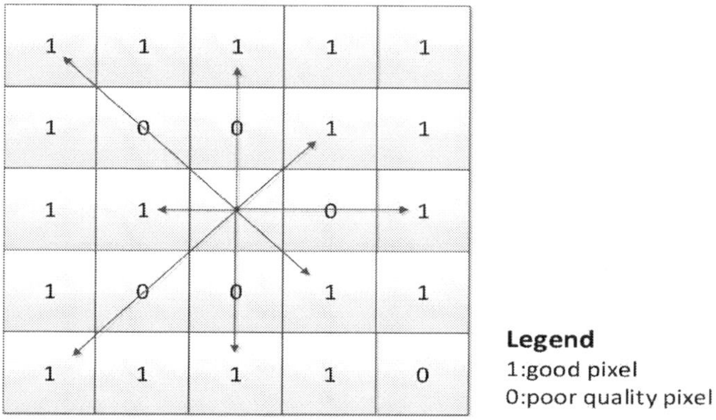


**Figure 2.** Eight directions searching.

Each pixel in the 163 images in seven years was studied using time series analyzing methods. Based on the EVI0 data, we firstly generated a time series curve of each pixel and dealt with noisy signal for each curve using Savitzky-Golay filter, and then extracted a long-term change trend (EVItr). After that, the weight of each point, which was in the 163 samples time series ($W_i$) data, was computed and the EVI time series were traversed.

### *1.1.2. Ecosystem (land cover): land cover semi-automatic classification from multispectral remote sensing imagery*

Through the conversion of forests and grasslands to croplands and pastures, humans have affected the exchange of energy, water and carbon between the atmosphere and the land surface. As the key input to ecosystem researches, various studies focused on land cover mapping [11]. Whereas, land cover data at large scale is hard to approached except remote sensing [12]. A variety of approaches was used to map land cover based on remote sensing data [13–17], including visual interpretation classification, unsupervised clustering coupled with extensive ancillary data and manual labeling of clusters, supervised classification, expert system classification, artificial intelligence neural network classification, and decision tree classification. However, the accuracy and the efficiency of land cover classification is not guaranteed and the land cover classifications are arbitrary using these method. For example, supervised classification methods require selecting training samples which rely on substantial expertise and human participation, so the result of land cover classification is influenced greatly, and it is impossible to classify land cover automatically with these methods. The algorithms such as neural network classification and fuzzy logic classification are difficult to understand and apply widely because of highly complicated in their algorithm basis. The construction of the decision tree and the assignment of thresholds for each sub-nodes are the key problem of decision tree classification, and they heavily depends on human experience and varies spatially and temporally. In order to solve these problems to improve the accuracy and efficiency of classification, Jiang et al. proposed an efficient automatic landscape classification approach using prior accurate land-cover data as the background experience [18, 19]. This method consists of two steps: (1) semi-automatically detecting land cover changed pixels from satellite images compared with prior land cover map; (2) semi-automatically classifying the land cover of changed pixels based on pattern recognition and changed rules.

**Automatic collection of training samples:** In this method, pure pixels of land cover were extracted automatically with an accurate previous land cover dataset as prior knowledge. The interiors of individual land cover areas and larger patches are considered to be more ecologically stabile areas. Based on accumulation area threshold (Pa), the samples of different land covers sorted in descending order of their area. The accumulation area threshold is calculated based on the percent of largest patches of specific land cover categories occupied in its total area.

After spatial buffer analysis, the joint region of different land cover were discarded. The buffer area are obtained using the different distance to all patches because of patches of vary areas. The distance of buffer analysis is

$$P_{buffer} = \frac{Ab_d}{A}, \quad d < 0 \tag{3}$$

where $P_{buffer}$ is area threshold for buffer analysis, $Ab_d$ is the buffer area of the patch with a distance of d, with d negative, and A is the area of the patch.

Establishment of three-dimensional feature space: The data in all spectral bands of each land cover class, which were extracted from the region of interest, were processed by principal component statistical analysis. The first three principal components for orthogonal decomposition was selected to construct the three-dimensional feature space of different land cover classes.

$$\sum_{j=1}^{3} \frac{\left(P_{ji} - MP_{ji}\right)^2}{\sigma_{ji}^2} < c_i^2 \tag{4}$$

where $P$ is the principal component (PC), MP and $\sigma$ are mean and standard deviation of PC, respectively, $c$ is the radius of the three-dimensional feature space.

Change detection and classification of changed pixels: Combining the satellite images and early land cover maps, the spectral data of the images were extracted according to accurate early land cover maps. Based on spatial intersect analysis with corresponding three-dimensional feature space, the pixels with spectral data outside the ellipsoid were detected as changed pixels. After obtaining the changed land cover pixels, the satellites images and three-dimensional feature space were used to classify the changed area of land cover by calculating the minimum spectral distance. For each changed pixel, the spectral data of all bands were input to the formula for all land cover classes in three-dimensional feature space to calculate the minimum spectral distance $d_{mi}$.

To express the rules of changed land cover, we proposed a drag coefficient of changed land cover (r). Combining $d_{mi}$ and r, we determined the final land cover classification of changed pixels. The minimum distance of the land cover classification based on changed rules is defined as:

$$dL_{mi} = p_{ij} * d_{mi} \tag{5}$$

where $dL_{mi}$ is the minimum distance of the $m$th pixel to the $i$th land cover class based on the changed rules and $r_{ij}$ is the drag coefficient of the $i$th land cover class that changed to the $j$th land cover class.

The class information would be contaminated by adjacent class codes, so the post-classification results was modified to solve the problem. Any group of pixels, which was <3 pixels in size, was identified as noise in dilation operation.

### *1.1.3. Energy parameter: land surface temperature*

The parameter of temperature of surface energy balances has be important index for studies of ecosystem response to climate change. As land cover mapping, the land surface temperature retrieval based on remote sensing has been the primary method for gaining temperature at large scale. This is due to the accessibility and convenience of satellite remote sensed information. However, the surface energy parameters retrieved from remote-sensing data are often interrupted on the spatial and temporal scales by clouds, aerosols, solar elevation angle and bidirectional reflection. so, it is always difficult to obtain complete dataset for a large region. In addition, the accuracy of surface parameter retrieval was affected with various degrees by an indirect retrieval method and the instantaneous features of monitoring [2, 20]. To reduce such impacts, a time-composite method is generally adopted. For surface energy parameters, the time series fitting and noise-removal methods commonly used to include the mean diurnal variation and nonlinear regression methods. Compared with nonlinear regression method, in-situ measurement data are necessary for the mean diurnal variation method [10, 21]. However, the modeled results often cannot represent the actual situation under significant changes of environmental conditions [20, 22]. In recent years, data-assimilation methods have been adopted in the reconstruction of time series data to solve these problems. In this paper, a time series reconstruction method based on an ensemble Kalman filter was established, which is can be applied in various studies[19, 23–26]. It focused on evaluating the state of a discrete-time controlled process. The state $X_k$ can be expressed as [27]:

$$X_k = \phi_{k,k-1} X_{k-1} + \Gamma_{k,k-1} W_{k-1} \tag{6}$$

where $Wk-1$ are noise in normal probability distributions; $\phi_{k,k-1}$ and $\Gamma_{k,k-1}$ are the noise covariance of process and measurement, respectively.

Further assumption was defined as:

$$\begin{cases} E[W_k] = 0, E[W_k W_j^T] = Q_k \delta_{kj} \\ E[V_k] = 0, E[V_k V_j^T] = R_k \delta_{kj} \\ E[W V_j^T] = 0 \end{cases} \tag{7}$$

where $Q_k$ is the nonnegative covariance matrix of $W_k$ and $R_k$ is the positive covariance matrix of $V_k$; $\delta_{kj}$ is the function of Kronecker-δ.

We define N as the number of days of measurement, then:

$$Q_k = 0.5 * \text{cov}(\text{randn}(1, N)) \tag{8}$$

$$R_k = 0.5 * \text{cov}(\text{randn}(1, N)) \tag{9}$$

where $\hat{X}_{k-1}$ is defined to be our a priori state estimate at step k, given knowledge of the process prior to the step k, and $\hat{X}_k$ is our a posteriori state estimate at step k, given measurement $Z_k$. Then, the a priori state is defined as follows:

$$\hat{X}_{k,k-1} = \phi_{k,k-1} \hat{X}_{k-1} \tag{10}$$

and the a posteriori state as follows:

$$\hat{X}_k = \hat{X}_{k,k-1} + K_k \left[ Z_k - H_k \hat{X}_{k,k-1} \right] \tag{11}$$

The calculation process of a Kalman filter is a constant "forecast correction" process. Based on time-update and observation-update, the values were reconstructed with minimum variance by compared to in situ values.

#### *1.1.4. Water balance: precipitation interpolation based on thin plate smoothing splines*

Precipitation data are one of the important input elements of ecological mechanism model, and it play an extremely important role in simulating and researching ecosystem changes at regional or global scale. It also a key index in global change researches, which is a direct parameter for drought estimation [28]. The precision of precipitation data may affect the simulation and prediction precision. At present, the precipitation data acquisition mainly rely on the long-term observation of the weather stations, but scarcity and uneven distribution of weather stations reduce the accurate of spatial precipitation data.

The precipitation measurements from the WMO stations are point measurements. Spatial interpolation technique for meteorological elements is often used to obtain the meteorological data of every position in the scope which is covered by the meteorological stations. Several methods have been developed to interpolate these point data to a real estimation of rainfall. The software ANUSPLIN, developed by Hutchinson, Australian National University, was applied to generate gridded precipitation data [29]. ANUSPLIN is a kind of analysis and interpolation tool for multivariable data using local thin plate spline function. The model required to enter the location of the meteorological site, elevation and other ancillary data. Its statistical analyzes and diagnose multivariable data to realize spatial interpolation function.

In order to fit to datasets distributed across an unlimited number of climate station locations, interpolating methods of ANUSPLIN were applied, which use thin plate smoothing splines for spatial interpolation with a third parameter of elevation [30–36]. The main advantage of thin plate splines is that splines do not require prior estimation of spatial auto-covariance structure, which was difficult to estimate and validate. The partial spline observational model for n data values $z_i$ at positions x is given by setting [37]:

$$Z_i = f(x_i)\sum_{j=1}^{p}\beta_j\varphi_j(x_i) + \varepsilon_i (i = 1,\ldots,n; j = 1,\ldots p) \tag{12}$$

where $f$ is the thin smoothing spline, $\beta_j$ are $j$th unknown parameter and $\varphi_j$ are $j$th known function, which all have to be estimated. $x_i$ is spatial position with elevation, $\varepsilon_i$ are errors with covariance structure given by

$$E\left(\varepsilon\varepsilon^T\right) = V\sigma^2 \tag{13}$$

where $\varepsilon = \left(\varepsilon, \ldots\varepsilon_m\right)^T$, $V$ is positive definite $n \times n$ matrix and $\sigma^2$ may be known or unknown, the errors $\varepsilon_i$ are uncorrelated if $V$ is diagonal and correlated otherwise. The function $f$ and the parameters $\beta$ are estimated by minimizing

$$\left(z-g\right)^T V^{-1}(z-g) + pJ_m(f) \tag{14}$$

where $z = \left(z, \ldots z_m\right)^T$and $g = \left(g, \ldots g_m\right)^T$with

$$g_i = g(x_i) = f(x_i) + \sum_{j=1}^{p}\beta_j\varphi_j(x_i) \tag{15}$$

where $J_m(f)$ is a measure of the roughness of the spline function $f$ defined in terms of $m$th order derivatives of $f$, and $p$ is a positive number called the smoothing parameter

### 1.2. Time series analysis

#### *1.2.1. Land cover change*

Land cover change is one of the main methods by which the human activities have effect on the land surface environment. Research on dynamic models of land use change process is an important approach and means to deeply understand land use change process and its causes, and the research also can reveal the response of the land cover to anthropogenic activities. Land cover change mainly shows in land cover change speed and transfer direction, and comprehensive land use dynamic degree, single land use dynamic degree and transition matrix were used to express them [38].

## 2. Comprehensive land cover dynamic degree

Comprehensive land use dynamic degree is used to describe regional difference of land use types change speed and reflect the influence on change of land cover types by human activities. The mathematical model are as follows:

$$S = \left[\sum_{i=1}^{n}\left(\Delta S_{i-j}/S_i\right)\right] \times 100 \Big/ t \times 100\% \qquad (16)$$

where $S$ is the comprehensive land cover dynamic degree in t period, $\Delta S_{i-j}$ is the total area of land cover change of $i$th type converted into $j$th type from monitor beginning to end. $S_i$ is the area of $i$th type when the monitor started.

## 3. Single land cover dynamic degree

Single land use dynamic degree is used to describe the speed and amplitude of different land cover types change in a certain period. It reflects the influence on change of single land cover type by human activities. The mathematical model is as follows:

$$K_i = \left(S_{it2} - S_{it1}\right)/S_{it1}/\left(t_2 t_1\right) \times 100\% \qquad (17)$$

where $K_i$ is the single land cover dynamic degree of $i$th type from t1 to t2 time, $S_{it2}$ and $S_{it1}$ were areas of $i$th type in $t_2$ and $t_1$ time separately.

## 4. Transition probability matrix

Transition probability matrix was proposed by the Russian mathematician Markov. At the beginning of the twentieth century, Markov found that the *n*th result affected by the *n-1*th result in the transfer of some factors of a system. In Markov's analysis, the quantitative descriptions of the system state and state transition were reflected in the transform process of a metastable system from time T to time T + 1 in a certain time interval, thus revealing the land cover pattern time and space evolution process. Transition matrix of land cover depicts comprehensively and specifically structural characteristics of land cover change and reflects the change direction led by human activities. Transition matrix are as follows:

$$S_{ij} = \begin{bmatrix} S_{11} & S_{12} & \cdots & S_{1n} \\ S_{21} & S_{22} & \cdots & S_{2n} \\ \cdots & \cdots & \cdots & \cdots \\ S_{n1} & S_{n2} & \cdots & S_{nn} \end{bmatrix} \tag{18}$$

### 4.1. Vegetation succession

Net primary production (NPP) represents the accumulated organic matter by plants per unit area and time. From an ecological perspective, it measures the rate at which solar energy is stored by plants as organic matter. NPP is influenced by climate, soil, vegetation type and human activities, for various ecological monitoring activities, and is generally regarded as an important factor that provides a comprehensive evaluation of ecosystem status and services, including productivity capability, habitat, and wildlife, and ecological footprint [39, 40]. NPP is not a directly observable ecosystem characteristic, and it is difficult to measure accurately over large areas due to the spatial variability of environmental conditions. A number of NPP models for different ecosystems have been developed. These models are broadly classified into regression-based and process-based. Regression-based models are established by empirically derived relationships between climate values and NPP, such as Miami [41]. Although regression-based models, with the advantages of simplicity and fewer parameter requirements, can be extrapolated for most land ecosystems, uncertainties are also involved when considering heterogeneous vegetation, standard errors of measurements and novel climatic conditions, which may not be appropriate for the regressions [42, 43]. Process-based models, ranging from simple models based on light use efficiency (LUE) to more mechanistic models based on "soil-vegetation-atmospheric-transfer" (SVAT) schemes, are based on physiological and ecological processes such as photosynthesis, evapotranspiration, respiration and nutrient cycling [44, 45]. These models have more parameter requirements and complexities; however, they better describe mechanisms and have the potential to estimate NPP more accurately when compared with regression-based models. The models based on LUE are called production efficiency models (PEMs), which use LUE for the conversion of absorbed photosynthetically active radiation (APAR) to biomass [46]. They are widely acceptable to map NPP at different scales as it follows the basic principles of the photosynthesis process and is easily amenable to remote sensing data [47]. The satellite data-driven PEMs, such as CASA [48] and GLO-PEM [49], have been used to analyze the spatiotemporal patterns of NPP over continents and global land surfaces [50–53].

The CASA model simulates NPP directly thus avoiding a Ra (autotrophic plant respiration) calculation and taking environmental conditions (temperature, rainfall/soil moisture) and vegetation characteristics into consideration [54, 55].

The CASA model computes NPP as a function of absorbed photo synthetically active radiation (APAR) and light use efficiency (LUE) [48, 56] as follows:

$$NPP(x,t) = APAR(x,t) \times LUE(x,t) \tag{19}$$

where $x$ represents the grid cell and t represents the period in which NPP is accumulated, for example, a month. APAR is determined by the fraction of photo synthetically active radiation (FPAR) and the total solar surface radiation (SOL) (MJ $m^{-2}$) [57] as

$$APAR(x,t) = SOL(x,t) \times FPAR(x,t) \times 0.5 \tag{20}$$

where the constant 0.5 represents the ratio of the total solar radiation (with a wavelength range of 0.4–0.7 μm) used by the vegetation [58].

LUE is calculated as the product of maximum light use efficiency, and its temperature and moisture stressors [56] as

$$LUE(x,t) = T_{\varepsilon 1}(x,t) \times T_{\varepsilon 2}(x,t) \times W_{\varepsilon}(x,t) \times \varepsilon_{\max} \tag{21}$$

where $LUE(x, t)$ represents the actual light use efficiency, $\varepsilon_{max}$ the maximum light use efficiency, and the value for grass (0.604 g/MJ), simulated by Running based on BIOME-BGC model [59], was used here; $T_{\varepsilon 1}(x, t)$ and $T_{\varepsilon 2}(x, t)$ are temperature scalars, and $W_{\varepsilon}(x, t)$ is the moisture stress coefficient. $T_{\varepsilon 1}(x, t)$, $T_{\varepsilon 2}(x, t)$ and $W_{\varepsilon}(x, t)$ were computed at every location at each time step. $T_{\varepsilon 1}(x, t)$ and $T_{\varepsilon 2}(x, t)$ are calculated as [56, 57].

$$T_{\varepsilon 1}(x) = 0.8 + 0.02 \times T_{opt}(x) - 0.0005 \times \left[T_{opt}(x)\right]_2 \tag{22}$$

$$T_{\varepsilon 2}(x,t) = \frac{1.1814}{\left\{1 + \exp\left[0.2 \times (T_{opt}(x) - 10 - T(x,t)\right]\right\} \times \left\{1 + \exp\left[0.3 \times \left(-T_{opt}(x) - 10 - T(x,t)\right)\right]\right\}} \tag{23}$$

where $T_{opt}$ is an optimal temperature, defined as the mean temperature in the month of maximum normal differential vegetation index (NDVI). T is the monthly mean temperature; $T_{\varepsilon 2}(x) = 1$, when $T = Topt$; it decreases to 0.5 when T is 10°C above or 13°C below $T_{opt}$.

$W_{\varepsilon}(x, t)$ reflects the effect of water condition, and it generally increases when available water increases. Atmospheric vapor pressure deficit reflects air humidity, which affects transpiration and then the LUE [60]. Therefore, there are currently studies using vapor pressure deficit (D in kPa) to calculate the moisture stress coefficient [61, 62], computed as [49]

$$W_{\varepsilon}(x,t) = (1.2e^{(-0.35D)}) - 0.2,$$

$$D = 0.611 \times \left[ \exp\left( (17.27 \times \frac{T_s - 273}{T_s - 36}) \right) - \exp\left( (17.27 \times \frac{T_d - 273}{T_d - 36}) \right) \right] \tag{24}$$

where $T_d$ is dew point temperature (K) and $T_s$ is surface temperature (K). When $T_s - T_d < 0, D = 0$. $T_d$ was derived from Guo Jie's regression model for Sichuan province based on Yang Jingmei's findings of a significant linear relationship between dew point temperature and the logarithm of total perceptible water [63, 64] as follows:

$$Ln(U) = 1.8084 + 0.0735T_d \tag{25}$$

where U is total perceptible water (mm).

### 4.2. Estimation of droughts: drought index, drought level definition, index of drought frequency

Drought is an kind of extreme water deficit processes, which can be used as indicators of ecosystem deteriorate and climate change. Drought can be classified to four categories of Meteorological Drought, Hydrological Drought, Agricultural Drought and Socioeconomic Drought [65–68]. There are numerous drought indices were formulated by integrating variables to identify and quantify the duration, magnitude, intensity and spatial extent of a drought, such as precipitation, evapotranspiration, temperature, terrestrial water storage (TWS), the TWS anomaly index (TWSI), vegetation, etc.[69–71]. Whereas, precipitation is the most direct parameter for evaluating meteorological droughts, which also can be applied in estimating agricultural drought, hydrological drought and socioeconomic drought [72–73]. To be a worldwide natural hazard, meteorological droughts can be measured by various indices such as Precipitation Anomaly Index (PAI) [69, 74, 75], Palmer Drought Severity Index(PDSI) [76], Z-score or standardized rainfall anomalies [77], Standardized Precipitation Index(SPI), Standardized Precipitation Evapotranspiration Index(SPEI) [78], et al. However, compare with complex indices, a simple measure may be applied more easily to evaluate drought disaster at large scale [79]. For example, PDSI requires, in addition to precipitation, soil moisture, runoff, evapotranspiration, potential evapotranspiration and other factors of plant growth to assess droughts. Furthermore, PDSI cannot be used to identify drought at short time scales, e.g., less than nine months [76]. Compared to complex indices, such as PDSI, SPI maybe a better index based on precipitation alone, as it also compares drought conditions among different time periods and regions. Among these indices, precipitation anomaly index (PAI) that uses precipitation alone is the simplest index; it is a dimensionless number in which negative/positive values indicate dry/wet conditions. It is precisely because of these advantages of simple computation, spatiotemporal consistency, and easy comparison to historical records, PAI is an important meteorological drought index for large area drought assessment in China [80].

The meteorological droughts index of the precipitation anomaly index (PAI) was used for drought analysis. PAI was calculated from the monthly precipitation data from the China meteorological data sharing service system (CMDSSS) of the China meteorological administration (CMA). The PAI of SC from 1961 to 2012 is calculated as:

$$PAI = (P - \bar{P}) / \bar{P} \times 100\% \tag{26}$$

where $P$ and $\bar{P}$ are precipitation and mean value.

| Droughts level | PAI (%) | | |
|---|---|---|---|
| | Month | Season | Year |
| None | [-40, ∞) | [-25, ∞) | [-15, ∞) |
| Mild | [-60, -40) | [-50, -25) | [-30, -15) |
| Moderate | [-80, -60) | [-70, -50) | [-40, -30) |
| Severe | [-95, -80) | [-80, -70) | [-45, -40) |
| Extreme | (- ∞, -95) | (- ∞, -80) | (- ∞, -45) |

**Table 1.** Drought level definition based on PAI.

Because of difficulty to evaluate absolute droughts, the meteorological drought level was used to estimate the drought severity. **Table 1** shows the relationship between the PAI and meteorological drought level in three temporal scales of month season and year [80].

The drought frequency (DF) were defined as:

$$DF = n/N \times 100\% \tag{27}$$

where n is the number of years of droughts, N is number of study years [81].

### 4.3. Correlate analysis between energy-water balance and ecosystem parameters

#### *4.3.1. Correlation coefficient*

Correlation coefficient method is used for studying the closeness relations of variables and is described by a quantitative index which is called correlation coefficient. The calculation process is simple and clear and the result is intuitional and easy to interpret, so the method is considered to be the best for analyzing long-term vegetation trends. When we study the interrelation of a plurality of geographic features, and study the impact of certain factor on the other feature without taking into account of other features, the correlation coefficient can be used. We chose correlation coefficient as the quantitative indicator of evaluation relevant. The formula of correlation coefficient is as follows [82]:

$$r_{xy} = \frac{\sum_{i=1}^{n}\left[(X_i - \overline{X})(Y_i - \overline{Y})\right]}{\sqrt{\sum_{i=1}^{n}(X_i - \overline{X})^2 \sum_{i=1}^{n}(Y_i - \overline{Y})^2}} \tag{28}$$

where $r_{xy}$ is the correlation coefficient of variable X and variable Y; n is the number of sample; $\overline{X}$ is the mean of variable X; $\overline{Y}$ is the mean of variable Y.

The significance test of correlation coefficient generally uses the t-test. The statistic calculation formula is:

$$r_{xy} = \frac{r\sqrt{n-2}}{\sqrt{1-r^2}} \tag{29}$$

where ***r*** is the correlation coefficient, ***n*** is the number of samples.

#### *4.3.2. Principal component analysis*

Principal component analysis (PCA) is a kind of multivariate statistical method. Through orthogonal transformation converts, a set of possible correlation between variables convert to a set of linear irrelevant variables which is called principal components.

Principal component analysis is to investigate correlation among multiple variables. It was used to study that how to reveal the internal structure among multiple variables by a few principal components. Namely, a few principal components were derived from the original variables and make them as much as possible to retain the information of the original variables and unrelated to one another. Usually, in mathematical treatment, the linear combination of the original index was regard as a new composite indicator. The classical approach is to use the variance of F1 (first linear combination) to express. The bigger Var(F1) is, the more information you gather F1 contains. Thus, the variance of F1 should be the biggest of all linear combinations and F1 is the first principal component. The number of principal components is decided by the quantity information that principal components represented [83].

## 5. Result and discussion

### 5.1. Climate change pattern of Southwest China

#### *5.1.1. Land surface temperature*

Taking Sichuan province as an example, the temperature increased from 1982 to 2010 (**Figure 3**). In regional terms, the temperature increased in 90.2% of the region in Sichuan province, in addition to Batang County, Derong County and Xiangcheng County. In western Sichuan region, Annual average temperature showed a rising trend, and it is very obvious in Jiuzhaigou County, Li County, Mili Tibetan Autonomous County, Shiqu County and Leibo County, the biggest increase among them is 1.89°C every 10 years. In eastern Sichuan, including Guangyuan, Bazhong, Nanchong, Ziyang, Leshan, Luzhong and so on, temperature increased by over 0.2°C every 10 years. The temperature rising trend in 76% of the region of the Sichuan province are extremely significant or significant. The temperatures have a decreased of as much as 1.25°C every 10 years in Batang County, Derong County and Xiangcheng County.

**Figure 4** shows the change of mean temperature of Sichuan Province in four seasons from 1982 to 2010. It turns out the spatial distribution pattern changes of mean temperature at different seasons are not significant. The changes of mean temperature in summer ranged from −6.59 to 22.06°C, the change ranged from −6.59 to 22.06°C in spring, from −6.59 to 22.06°C in autumn, and from −6.59 to 22.06°C in winter. The mean temperature variations in winner is biggest (30.746°C), and it in autumn is smallest (25.27°C).

**Figure 5** shows the monthly mean temperature changes of time series from 1982 to 2010. As can be seen from the **Figure 5**, the monthly mean temperature display cyclic variations under this time series. It offered upgrade firstly than descending latter tendency. In June, July, August and September, the temperature was higher and reaches a maximum in July and August. In January, the monthly mean temperature is lower. The changes of every year were basically similar, and they have certain periodicity with an obvious fluctuation. The fluctuation of the highest temperature of each year is very small. A maximum of the highest temperature appears in the June 2006 was 23.65°CA minimum of the highest temperature has an obvious fluctuation ranged from 1.23(1984) to 4.28(1987).

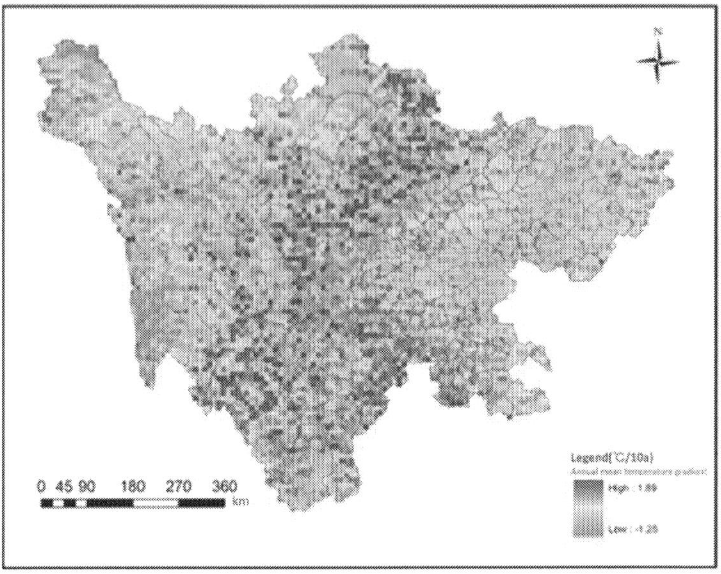


Figure 3. Gradient of the annual mean temperature of Sichuan Province.

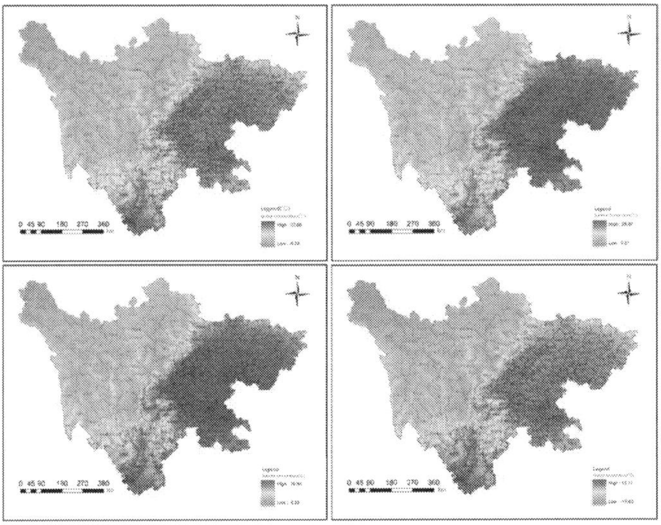

Figure 4. Different seasonal mean distribution maps of Sichuan Province.

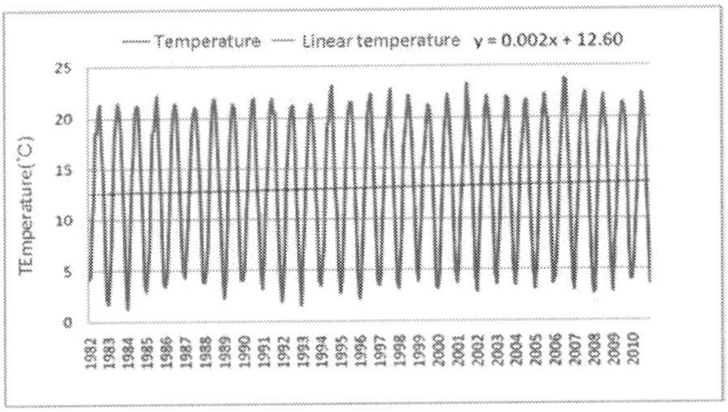


**Figure 5.** Change curve of monthly mean temperature during 1982–2010.

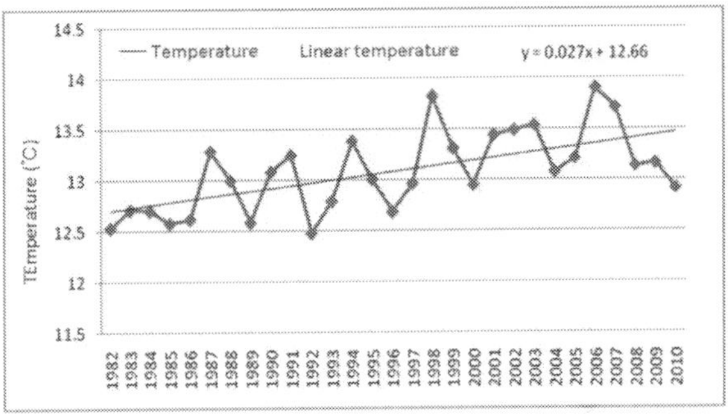


**Figure 6.** Change curve of annual mean temperature during 1982–2010.

**Figure 6** shows annual changes of the mean temperature from 1982 to 2010. The mean temperature fluctuated between 12.48 and 13.90°C. The minimum of the mean temperature occurred in 1992 and the maximum occurred in 2006. From 1982 to 2010, the annual mean temperature was rising obviously which was consistent with global warming trends.

### *5.1.2. Precipitation*

Taking Sichuan province as an example, we selected the monthly mean precipitation data of nine weather stations from 1982 to 2010. As can be seen from **Figure 7**, the spatial distribution of annual average precipitation was uneven in 29 years with descending from the east to the west. Taking Songpan Country-Li Country-Kangding City-Jiulong Country-Yanyuan Country as a boundary, the annual precipitation of the east of the boundary was abundant, and it is more than 900 mm/a. Especially, the annual mean precipitation of Ya'an is more than 1200 mm/a and ranked first in the whole province. The annual mean precipitation of most areas in the west is not more than 800 mm. Among them, there is a very little rain in Shiqu country, less than 600 m/a.

As shown in **Figure 8**, the annual average precipitation was concentrated mostly in summer, and there are obvious differences in the spatial distributions in different seasons. The spring precipitation mainly concentrated in the central and eastern regions in Sichuan, and the maximum precipitation was 311.55 mm. The precipitation in summer moves westward concentrating in the central regions with the maximum precipitation of 859.68 mm. The scope of autumn rainfall mainly concentrated in the central and southern Sichuan with uneven spatial distribution. The maximum precipitation was 340.02 mm. The precipitation of most regions in winter was low, and decreased from the west to the east.

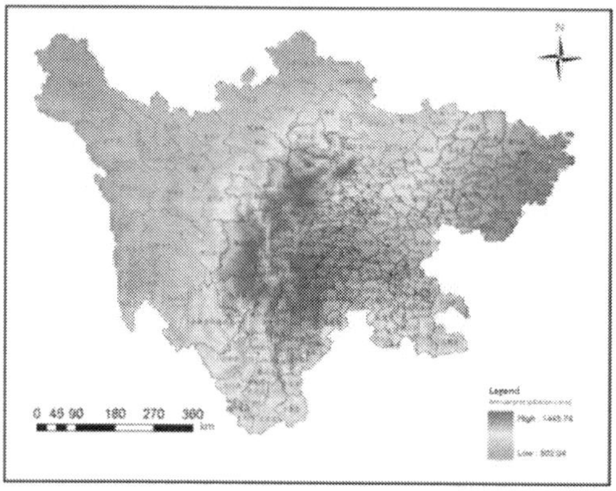


**Figure 7.** Annual precipitation distribution map of Sichuan Province.

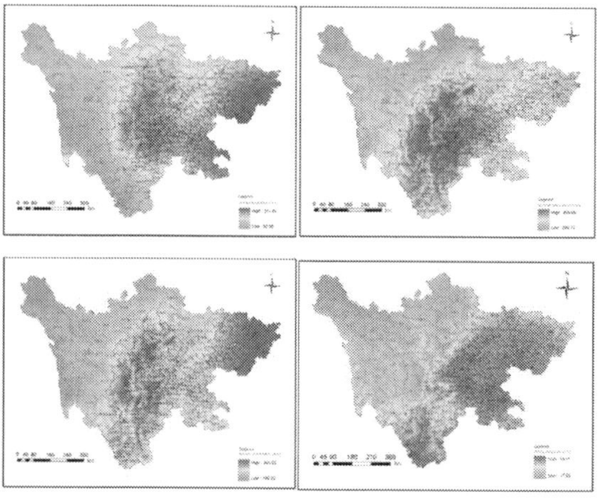

**Figure 8.** Different precipitation distribution maps of Sichuan Province in different seasons.

Monthly mean precipitation changing trends in every years were basically the same, namely, it offered upgrade firstly than descending latter tendency (**Figure 9**). In June, July, August and September, the precipitation was higher and reaches a maximum in July and August. In January, the monthly mean precipitation is lower. The precipitation change has certain periodicity with an obvious fluctuation from 1982 to 2010. The maximum of annual highest precipitation was 294.49 mm in 1984, and the minimum is 154.03 mm in 2006. The annual lowest precipitation has no obvious fluctuation ranged from 2.86 to 15.71 mm. The monthly mean precipitation had not a visible downtrend by 0.11 mm per decade.

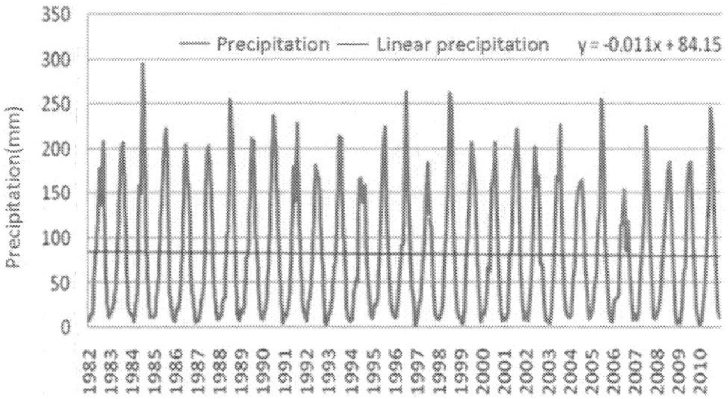


**Figure 9.** Change curve of monthly mean precipitation during 1982–2010.

**Figure 10** shows the changes of the annual precipitation from 1982 to 2010. The annual precipitation showed obvious decreasing trend. The maximum of the annual precipitation occurred in 1989 (1131.17 mm), and the minimum occurred in 2006 (812.84 mm). The precipitation changed in a wide range and had obvious fluctuations.

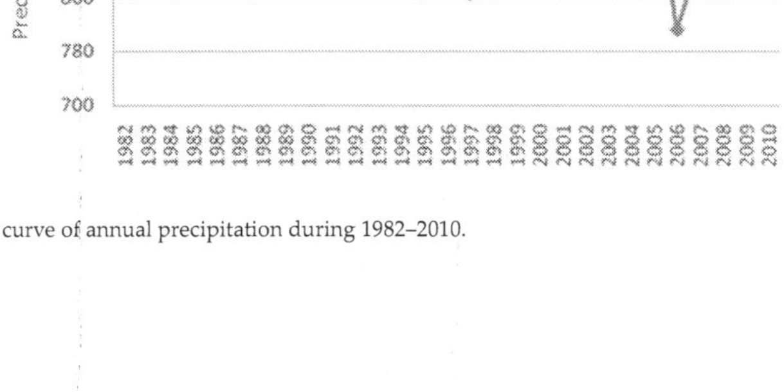


**Figure 10.** Change curve of annual precipitation during 1982–2010.

## 5.2. Variability of ecosystem in Southwest China

### *5.2.1. Land cover change*

We integrated remote sensing data and other correlative material to get the land use maps for 1980, 1990, 1995, 2000 and 2005 years (**Figure 11**). The maps included 18 land cover types: dry land, paddy fields, low coverage grassland, medium coverage grassland, high coverage grassland, sparse woodland, shrub land, woodland, other wood land, lakes, graff, reservoir and pits, coast, beach, urban land, rural residential areas, construction land, waste land. As can be seen from **Figure 11**, the main land use types were cropland, grassland, woodland. Among them, paddy field and dry land were highly concentrated in Sichuan and Chongqing and dry land accounted for a large proportion. There are only small numbers of them scattered across the other region. The grassland, including low coverage grassland, medium coverage grassland and high coverage grassland, were distributed mainly in Sichuan and Guizhou. The woodland were mainly distributed at Yunnan and Guangxi.

The time series analysis on land cover indicated the whole change is not obvious with certain changes on land use types during 20 years. High coverage grassland, woodland, urban land and construction land increase, others decrease.

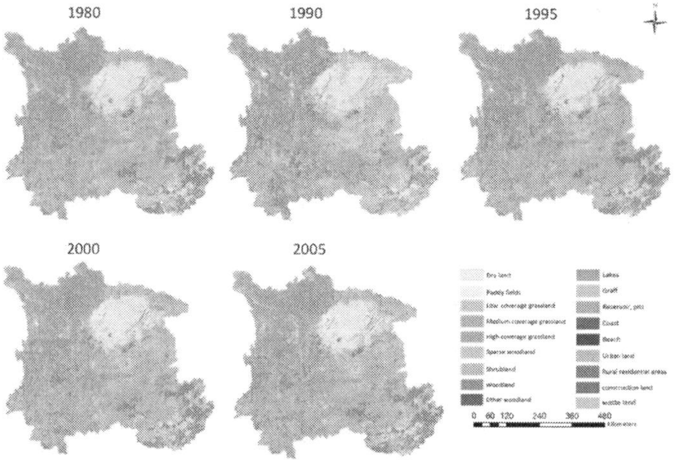


**Figure 11.** Distribution of land cover in 1980, 1990, 1995, 2000 and 2005.

### *5.2.2. Vegetation destruction and recovery*

The normal differential vegetation index (NDVI) products MOD13A2 (d209-d225) were obtained from the NASA website (ftp://e4ftl01.cr.usgs.gov/) at the same time from 2004 to 2010. These data were used to analyze the vegetation destruction and recovery. As can be seen from

**Figure 12**, NDVI is smaller in the northwest of Sichuan Province than other area in Southwest China. Comparing 2006 with 2004, NDVI in southeast of Yunnan and Chongqing Province changed significantly with a sudden drop, and it in other areas showed no obvious change. In 2008, the conditions had improved, NDVI in southeast of Yunnan and Chongqing Province increased, but it in the northwest of Sichuan Province decreased. In addition, there is a large scale of the decline of NDVI in the south-central part of Southwest China. NDVI in some regions of northwest of Sichuan Province increased in 2010, and it decreased in the region near the boundary of Sichuan and Chongqing Province. In general, NDVI in Southwest China decreased from 2004 to 2010, especially in Sichuan Province.

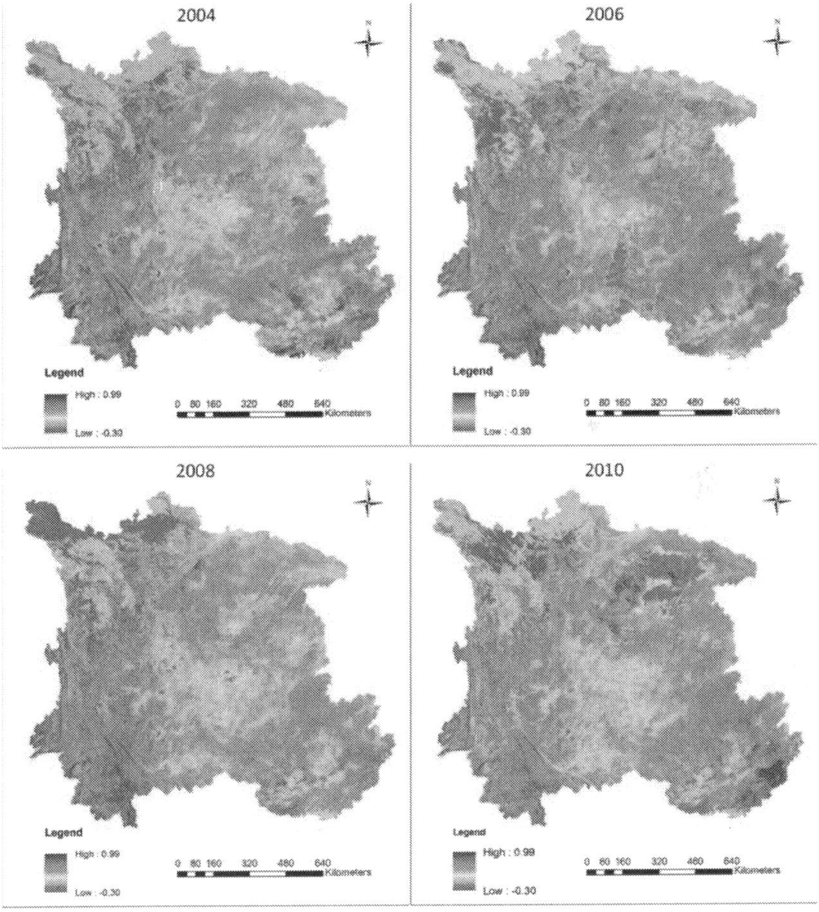


**Figure 12.** Distribution of NDVI in 1980, 1990, 1995, 2000 and 2005.

### 5.3. Energy-water balance and ecosystem response to climate change

#### *5.3.1. Energy-water balance response: droughts analysis based on precipitation*

The spatial distribution variability of the index of drought frequency (DF) of 12 months from 1961 to 2012 in SC is shown as **Figure 13**. Monthly DF has clear change in different months. SC suffered from droughts in a large area in January, February, March, October, November and December. As shown in **Figure 13**, Yunnan Province and Guangxi Province are severely drought-afflicted areas. A drought pattern also appeared in the monthly variability of DF over time. From January to March, central and eastern Yunnan Province, southwestern Sichuan Province, and southern Guangxi Province experienced droughts (DF >40%) over large area. In May, the drought area rapidly narrowed, and the area of DF >40% was located in a limited extension of Yunnan Province. From June to September, the DF was low with a slightly increase. The distribution of droughts was spread from east to west. In October, droughts began to occur again in a large area, and the DF of eastern Guangxi Province and the DF of northwestern

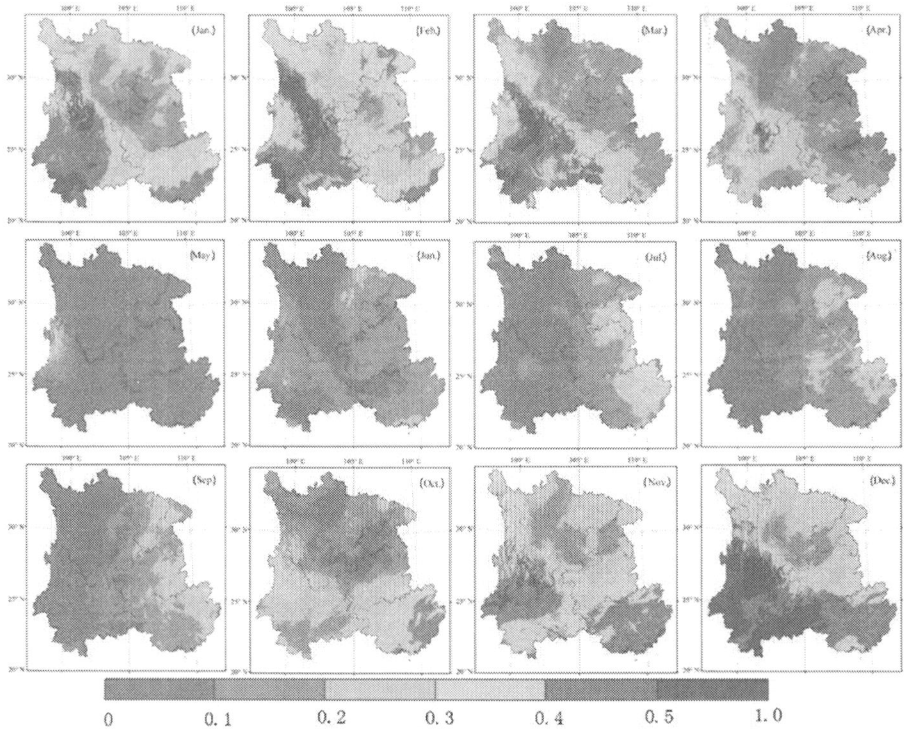


**Figure 13.** Spatial distribution of the monthly DF in SC from 1961 to 2012.

Yunnan Province exceeded 40 and 30%, respectively. From November, the drought area increased rapidly. Half of study area is at a high drought risk. The DF of the entire Yunnan Province in western SC is >40%. In December, the DF of most of the area of Yunnan Province exceeded 50%, except in the limited extension of the central portion, which meaning that these regions suffered droughts biyearly.

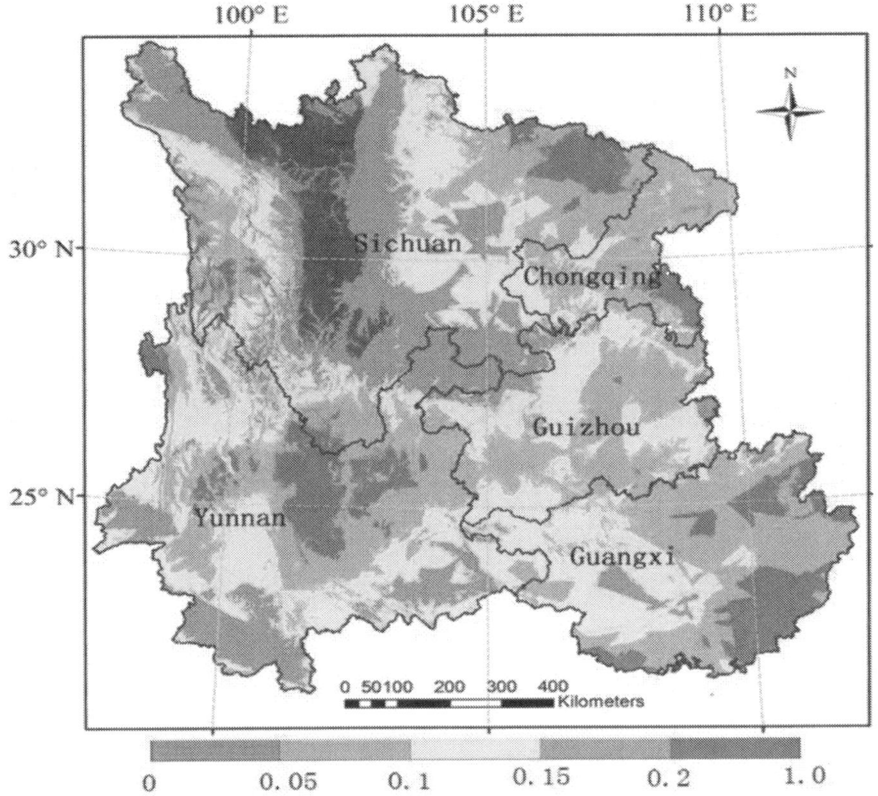


**Figure 14.** Spatial distribution of annual DF in Southwest China from 1961 to 2012.

To illustrate the spatial distribution of drought variability, the annual DF of SC from 1961 to 2012 was calculated. As shown in **Figure 14**, the annual droughts DF were scattered, in contrast to the monthly scale. More than 61% of the area in SC had a relatively high DF (>15%), that is, all of SC has been a region of high drought risk for more than half century. The southern and eastern mountain zones have high drought frequency. In contrast, Sichuan Basin, which occupies a large portion of the study area with relatively flat and fertile grounds, suffered fewer droughts events.

*5.3.2. Ecosystem response: droughts analysis based on vegetation index*

The vegetation index data were 1 km, 16 days composited MODIS EVI (MOD13A12), which were downloaded from the NASA EOS Data Gateway (EDG) (http://modis.gsfc.nasa.gov/index.php). To be consistent with monthly GRACE estimates, the days EVI product was recomposited using the maximum value composite (MVC). The composited monthly EVI data were then linearly interpolated to a 1 × 1 grid.

Droughts are common in Southwest China, and several episodes of potential severe droughts were detected using temporal variability analysis of the GRACE TWS change. To evaluate drought events in Southwest China from 2003 to 2013, three indicators were taken into account: the PAI, the AVI and the annual cycle removed TWS (TWSI). To compute the TWSI, monthly averaged GRACE TWS change data for 10 years were removed at each pixel. Using the spatiotemporal variability analysis described above for the TWS, the precipitation, and the EVI, we chose representative pixels from three regions for drought analysis. **Figure 15** shows the time series of the three droughts indicators between 2006 and 2012 when droughts were likely to occur in Southwest China.

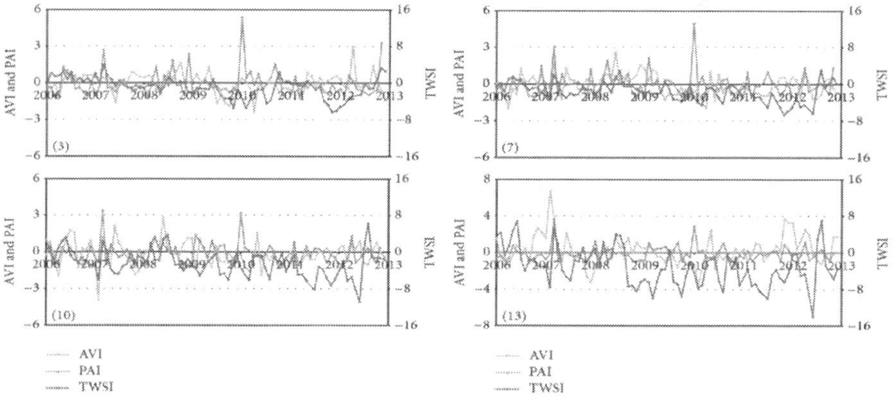


**Figure 15.** Time series of the TWSI, the PAI, and the AVI for selected pixels.

The correlation coefficients between EVI and precipitation were all more than 0.84 from 2003 to 2013, whereas they were 0.05, 0.07, 0.08 and 0.16 between PAI and AVI, respectively. The low correlation coefficients between PAI and AVI imply that these two drought indicators predict different water resources deficit conditions that accompany droughts. However, correlation analysis showed that TWSI present low correlation with EVI, precipitation and corresponding droughts indicators. **Figure 15** shows that the occurrence and release of drought AVI lagged behind PAI for 1–3 months and droughts of AVI were more severe than PAI. This is because of the delay in recharge of surface and soil water from rainfall and the vegetation growth. Both indicators mainly reflected water depletion in surface water and shallow soil water. PAI and AVI were both invalid for detecting droughts. This was most obvious during

http://dx.doi.org/10.5772/64960

the period from the end of 2011 to the beginning of 2012 in **Figure 15**, when the time series for PAI and AVI had no extreme droughts and a normal and gentle amplitude fluctuation. However, TWSI showed that Southwest China experienced great water resource decreases during this period, which may have caused an extreme drought in 2006. Considering the annual cycle of precipitation and vegetation growth and their relation to shallow water, the TWS change mainly contributed to the discharge of groundwater to surface water, which implied a drought risk.

## 6. Conclusion

Southwest China is an ecologically fragile area with more than 242 million populations. The change of water-energy balance and ecosystem due to climate change will affect the regional ecological and human living significantly. Our study of Southwest China in past 30 years shows that ecosystem structures destroyed by shrinking of water ecosystem, forest ecosystem and grass ecosystem, together with the precipitation reduction and temperature rise. In addition, the climate changes also affected the artificial ecosystems such as crops land, which resulting in food production mainly caused by frequently occurred severe droughts. Our research showed that, even in the region with abundant water resources of Southwest China, energy-water balance and ecosystem have severe response to climate change, which is significant to human productions and activities of daily livings.

## Author details

Jianhua Wang[1*], Yaohuan Hang[2], Dong Jiang[2] and Xiaoyang Song[3]

*Address all correspondence to: jiangd@igsnrr.ac.cn

1 State Key Laboratory of Simulation and Regulation of Water Cycle in River Basin, China Institute of Water Resources and Hydropower Research, Beijing, China

2 Institute of Geographical Sciences and Natural Resources Research, Chinese Academy of Sciences, Beijing, China

3 China University of Mining and Technology, Beijing, China

## References

[1] Stephenson N L. Climatic control of vegetation distribution—the role of the water-balance. American Naturalist, 1990, 135(5): 649–670.

[2] Francis A P, Currie D J. A globally consistent richness-climate relationship for angiosperms. American Naturalist, 2003, 161(4): 523–536.

[3] Zhao Y-S, Chen D-M, Yang L-M. The application principle and method of remote sensing. Science Press, Beijing, 2003.

[4] Tucker C J, Townshend J R G, Goff T E. African land-cover classification using satellite data. Science, 1985, 227(4685): 369–375.

[5] Justice C O, Townshend J R G, Holben B N, et al. Analysis of the phenology of global vegetation using meteorological satellite data. International Journal of Remote Sensing, 1985, 6(8): 1271–1318.

[6] Sakamoto T, Yokozawa M, Toritani H, et al. A crop phenology detection method using time-series MODIS data. Remote Sensing of Environment, 2005, 96(3–4): 366–374.

[7] Unganai L S, Kogan F N. Drought monitoring and corn yield estimation in Southern Africa from AVHRR data. Remote Sensing of Environment, 1998, 63(3): 219–232.

[8] Ma M G, Veroustraete F. Reconstructing pathfinder AVHRR land NDVI time-series data for the Northwest of China. Natural Hazards and Oceanographic Processes from Satellite Data, 2006, 37(4): 835–840.

[9] Viovy N, Arino O, Belward A S. The best index slope extraction (Bise)—a method for reducing noise in NDVI time-series. International Journal of Remote Sensing, 1992, 13(8): 1585–1590.

[10] Chen J, Jonsson P, Tamura M, et al. A simple method for reconstructing a high-quality NDVI time-series data set based on the Savitzky–Golay filter. Remote Sensing of Environment, 2004, 91(3–4): 332–344.

[11] Friedl M A, Brodley C E. Decision tree classification of land cover from remotely sensed data. Remote Sensing of Environment, 1997, 61(3): 399–409.

[12] Rogan J, Chen D M. Remote sensing technology for mapping and monitoring land-cover and land-use change. Progress in Planning, 2004, 61: 301–325.

[13] You-Shui Z, Xue-Zhi F, Ren-Zong R. Application of back-propagation neural network supported by GIS in the classification of remote sensing image. Journal-Nanjing University Natural Sciences Edition, 2003, 39(6): 806–813.

[14] Bao-Guang Z. Application of fuzzy mathematics to classification processing of remote sensing digital images. Journal of Tianjin Normal University (Natural Science Edition), 2005, 2: 020.

[15] Loveland T R, Merchant J W, Ohlen D O, et al. Development of a land-cover characteristics database for the conterminous United-States. Photogrammetric Engineering and Remote Sensing, 1991, 57(11): 1453–1463.

[16] Schneider A, Friedl M A, Potere D. Mapping global urban areas using MODIS 500-m data: new methods and datasets based on 'urban ecoregions'. Remote Sensing of Environment, 2010, 114(8): 1733–1746.

[17] Liu J Y, Liu M L, Tian H Q, et al. Spatial and temporal patterns of China's cropland during 1990–2000: an analysis based on Landsat TM data. Remote Sensing of Environment, 2005, 98(4): 442–456.

[18] Jiang D, Huang Y H, Zhuang D F, et al. A simple semi-automatic approach for land cover classification from multispectral remote sensing imagery. Plos One, 2012, 7(9):e45889-e45889.

[19] Fu J Y, Jiang D, Huang Y H, et al. A Kalman filter-based method for reconstructing GMS-5 global solar radiation by introduction of in situ data. Energies, 2013, 6(6): 2804–2818.

[20] Neteler M. Estimating daily land surface temperatures in mountainous environments by reconstructed MODIS LST data. Remote Sensing, 2010, 2(1): 333–351.

[21] Huang Y H, Wang J H, J D, et al. Reconstruction of MODIS-EVI time-series data with S-G filter. Geomatics and Information Science of Wuhan University, 2009, 34(12): 1440–1443.

[22] Ziwei X U, Liu S, Tongren X U, et al. Comparison of the gap filling methods of evapotranspiration measured by eddy covariance system. Advances in Earth Science, 2009, 24(4): 372–382.

[23] Aravequia J A, Szunyogh I, Fertig E J, et al. Evaluation of a strategy for the assimilation of satellite radiance observations with the local ensemble transform Kalman filter. Monthly Weather Review, 2011, 139(6): 1932–1951.

[24] Huang C L, Li X, Lu L. Retrieving soil temperature profile by assimilating MODIS LST products with ensemble Kalman filter. Remote Sensing of Environment, 2008, 112(4): 1320–1336.

[25] Rosema A. Using meteosat for operational evapotranspiration and biomass monitoring in the Sahel region. Remote Sensing of Environment, 1993, 46(1): 27–44.

[26] Sakov P, Evensen G, Bertino L. Asynchronous data assimilation with the EnKF. Tellus Series a-Dynamic Meteorology and Oceanography, 2010, 62(1): 24–29.

[27] Welch G, Bishop G. An introduction to the Kalman filter. University of North Carolina at Chapel Hill, Springer, Heidelberg, Berlin, 1995, vol. 785: 127–132. ISBN 978-3-642-02287-6.

[28] Wang J H, Jiang D, Huang Y H, et al. Drought analysis of the Haihe River Basin based on GRACE terrestrial water storage. Scientific World Journal, 2014, 2014:578372-578372.

[29] Huang Y, Xu C, Yang H, et al. Temporal and spatial variability of droughts in Southwest China from 1961 to 2012. Sustainability, 2015, 7(10): 13597–13609.

[30] Fleming M D, Chapin F S, Cramer W, et al. Geographic patterns and dynamics of Alaskan climate interpolated from a sparse station record. Global Change Biology, 2000, 6(S1): 49–58.

[31] He B, Lü A, Wu J, et al. Drought hazard assessment and spatial characteristics analysis in China. Journal of Geographical Sciences, 2011, 21(2): 235–249.

[32] Hong X, Guo S, Xiong L, et al. Spatial and temporal analysis of drought using entropy-based standardized precipitation index: a case study in Poyang Lake basin, China. Theoretical and Applied Climatology, 2015, 122(3–4): 543–556.

[33] Hutchinson M F. Interpolation of rainfall data with thin plate smoothing splines—Part I: two dimensional smoothing of data with short range correlation. Journal of Geographic Information and Decision Analysis, 1998, 2(2): 153–167.

[34] Liu Z, Xingxiu Y U, Wang S, et al. Comparative analysis of three covariates methods in thin-plate smoothing splines for interpolating precipitation. Progress in Geography, 2012, 31(1): 56–62.

[35] Price D T, Mckenney D W, Nalder I A, et al. A comparison of two statistical methods for spatial interpolation of Canadian monthly mean climate data. Agricultural and Forest Meteorology, 2000, 101(s 2–3): 81–94.

[36] Qian Y L, Lv H Q, Zhang Y H. Application and assessment of spatial interpolation method on daily meteorological elements based on ANUSPLIN software. Journal of Meteorology and Environment, 2010, 26(2):7-14.

[37] Hutchinson M F. Interpolation of rainfall data with thin plate smoothing splines—Part I: two dimensional smoothing of data with short range correlation. Journal of Geographic Information and Decision Analysis, 1998, 2(2): 139–151.

[38] Jiyuan L. Study on spatial-temporal feature of modern land use change in China: using remote sensing techniques. Quaternary Sciences, 2000, 20(3): 11.

[39] Crabtree R, Potter C, Mullen R, et al. A modeling and spatio-temporal analysis framework for monitoring environmental change using NPP as an ecosystem indicator. Remote Sensing of Environment, 2009, 113(7): 1486–1496.

[40] Ramakrishna N, Michael W, Peter T, et al. Recent trends in hydrologic balance have enhanced the carbon sink in the United States. Geophysical Research Letters, 2002, 29(10): 106-1–106-4.

[41] Lieth H. Modeling the primary productivity of the world. Springer, Heidelberg, Berlin, 1975: 237–263.

[42] Melillo J M, Mcguire A D, Kicklighter D W, et al. Global climate change and terrestrial net primary production. Nature International Weekly Journal of Science, 1993, 363(6426): 234–240.

[43] Golubyatnikov L L, Denisenko E A. Modeling the values of net primary production for the zonal vegetation of European Russia. Izvestiia Akademii Nauk Seriia Biologicheskaia, 2001, 28(3): 353–361.

[44] Wang P, Xie D, Zhou Y, et al. Estimation of net primary productivity using a process-based model in Gansu Province, Northwest China. Environmental Earth Sciences, 2014, 71(2): 647–658.

[45] Bala G, Joshi J, Chaturvedi R K, et al. Trends and variability of AVHRR-derived NPP in India. Remote Sensing, 2013, 5(2): 810–829.

[46] Cramer W, Kicklighter D W, Bondeau A, et al. Comparing global models of terrestrial net primary productivity (NPP): overview and key results. Global Change Biology, 1999, 5(S1): 1–15.

[47] Kale M P, Roy P S. Net primary productivity estimation and its relationship with tree diversity for tropical dry deciduous forests of central India. Biodiversity and Conservation, 2012, 21(5): 1199–1214.

[48] Potter C S, Rerson J T, Field C B, et al. Terrestrial ecosystem production: a process model based on global satellite and surface data. Global Biogeochemical Cycles, 1993, 7(4): 811–841.

[49] Prince S D, Goward S N. Global primary production: a remote sensing approach. Journal of Biogeography, 1995, 22(4/5): 815–835.

[50] Cao M, Prince S D, Small J, et al. Remotely sensed interannual variations and trends in terrestrial net primary productivity 1981–2000. Ecosystems, 2004, 7(3): 233–242.

[51] Nayak R K, Patel N R, Dadhwal V K. Estimation and analysis of terrestrial net primary productivity over India by remote-sensing-driven terrestrial biosphere model. Environmental Monitoring and Assessment, 2010, 170(1–4): 195–213.

[52] Potter C, Klooster S, Myneni R, et al. Continental-scale comparisons of terrestrial carbon sinks estimated from satellite data and ecosystem modeling 1982–1998. Global and Planetary Change, 2003, 39(s 3–4): 201–213.

[53] Zhang Y, Wei Q, Zhou C, et al. Spatial and temporal variability in the net primary production of alpine grassland on the Tibetan Plateau since 1982. Journal of Geographical Sciences, 2014, 24(2): 269–287.

[54] Bian J, Li A, Deng W. Estimation and analysis of net primary Productivity of Ruoergai Wetland in China for the recent 10 years based on remote sensing. Procedia Environmental Sciences, 2010, 2(6): 288–301.

[55] Mccallum I, Wagner W, Schmullius C, et al. Satellite-based terrestrial production efficiency modeling. Carbon Balance and Management, 2009, 4(9): 1–14.

[56] Field C B, Randerson J T, Malmström C M. Global net primary production: combining ecology and remote sensing. Remote Sensing of Environment, 1995, 51(1): 74–88.

[57] Piao S L, Fang J Y, Guo Q H. Application of casa model to the estimation of Chinese terrestrial net primary productivity. Acta Phytoecologica Sinica, 2001, 25:603-608

[58] Mccree K J. Photosynthetically active radiation. Springer, Heidelberg, Berlin, 1981: 41–55.

[59] Running S W, Thornton P E, Nemani R, et al. Global terrestrial gross and net primary productivity from the earth observing system. Springer, New York, 2000: 44–57.

[60] Jiang L P, Qin Z H, Xie W, et al. A research of net primary productivity model of grassland based on MODIS data. Chinese Journal of Grassland, 2006, 28(6): 72–76.

[61] Chen L J, Liu G H, Feng X F. Estimation of net primary productivity of terrestrial vegetation in China by remote sensing. Acta Botanica Sinica, 2001, 43(11): 1191–1198.

[62] Chen L J, Liu G H, Hui Guo L I. Estimating net primary productivity of terrestrial vegetation in China using remote sensing. Journal of Remote Sensing, 2002, 6(2): 129–135.

[63] Jie G, Ping L G. Climatic characteristics of precipitable water vapor and relations to surface water vapor column in Sichuan and Chongqing region. Journal of Natural Resources, 2009, 24(2): 344–350.

[64] Yang J, Qiu J. A method for estimating precipitable water and effective water vapor content from ground humidity parameters. Chinese Journal of Atmospheric Sciences, 2002, 26(1): 9–22.

[65] Keyantash J, Dracup J A. The quantification of drought: an evaluation of drought indices. Bulletin of the American Meteorological Society, 2002, 83(8): 1167–1180.

[66] Lu X, Zhuang Q. Areal changes of land ecosystems in the Alaskan Yukon River Basin from 1984 to 2008. Environmental Research Letters, 2011, 6(3): 329–346.

[67] Raziei T, Saghafian B, Paulo A A, et al. Spatial patterns and temporal variability of drought in Western Iran. Water Resources Management, 2009, 23(3): 439–455.

[68] Glantz D A W, H. M. Understanding: the drought phenomenon: the role of definitions. Water International, 1985, 10(3): 111–120.

[69] Wang J, Jiang D, Huang Y, et al. Drought analysis of the Haihe River Basin based on GRACE terrestrial water storage. Advances in Meteorology, 2014, 2014: 578372–578372.

[70] Lu X, Zhuang Q. Evaluating evapotranspiration and water-use efficiency of terrestrial ecosystems in the conterminous United States using MODIS and AmeriFlux data. Remote Sensing of Environment, 2010, 114(9): 1924–1939.

[71] Huang Y, Wang J, Dong J, et al. Surface water deficiency zoning of China based on surface water deficit index (SWDI). Water Resources, 2014, 41(4): 481–487.

[72] Vicente-Serrano S M. Differences in spatial patterns of drought on different time scales: an analysis of the Iberian Peninsula. Water Resources Management, 2006, 20(1): 37–60.

[73] Li S, Xiong L, Dong L, et al. Effects of the Three Gorges Reservoir on the hydrological droughts at the downstream Yichang station during 2003–2011. Hydrological Processes, 2013, 27(26): 3981–3993.

[74] Kisaka M O, Mucheru-Muna M, Ngetich F K, et al. Rainfall variability, drought characterization, and efficacy of rainfall data reconstruction: case of Eastern Kenya. Advances in Meteorology, 2014, 2014 (Article ID 380404): 1–6.

[75] Zhang Q, Li J, Singh V P, et al. Influence of ENSO on precipitation in the East River basin, South China. Journal of Geophysical Research Atmospheres, 2013, 118(5): 2207–2219.

[76] Guttman N B. Comparing the palmer drought index and the standardized precipitation index. Jawra Journal of the American Water Resources Association, 1998, 34(1): 113–121.

[77] Jones P D, Hulme M. Calculating regional climatic time series for temperature and precipitation: methods and illustrations. International Journal of Climatology, 1996, 16(4): 361–377.

[78] Vicente-Serrano S M, Beguería S, López-Moreno J I. A multiscalar drought index sensitive to global warming: the standardized precipitation evapotranspiration index. Journal of Climate, 2010, 23(7): 1696–1718.

[79] Oladipo E O. A comparative performance analysis of three meteorological drought indices. International Journal of Climatology, 1985, 5(6): 655–664.

[80] Zhang Q, Zou X, Xiao F, et al. Classification of meteorological drought. Standardization of Chinese (English), 2011: 52–55.

[81] Bao Y, Meng C, Shen S, et al. Temporal and spatial patterns of droughts for recent 50 years in Jiangsu based on meteorological drought composite index. Acta Geographica Sinica, 2011, 66(5): 599–608.

[82] Lachenbruch P. A relation between the correlation coefficient and the t, test using an indicator variable. Biometrische Zeitschrift, 1973, 15(4): 243–244.

[83] Jolliffe I T. Principal component analysis. Springer, Heidelberg, Berlin, 2010: 41–64.

Chapter 4

# Climate Parameter Variability and Health

Narges Khanjani

Additional information is available at the end of the chapter

http://dx.doi.org/10.5772/64955

**Abstract**

Over the past 50 years, human activities such as burning fossil fuels have released huge quantities of greenhouse gases, which have trapped additional heat in the lower layers of atmosphere, changed global climate and led to more intense and frequent weather events. The overall health effects of climate change are likely to be extremely negative. Climate change affects social and environmental factors related to health, such as drinking water, food and shelter. It also imposes new disease and mortality on human populations. Extreme high temperatures increase deaths from trauma, diabetes, mental disorders and cardiovascular, respiratory and renal disease. As the number of weather-related natural disasters increase every year, these disasters result in more deaths and slams the basic living need of people, mainly in developing countries. Intense rainfall and flood, ruin agricultural land, contaminate freshwater supplies, increase the risk of waterborne diseases, and create breeding grounds for disease-carrying insects and increase the incidence of infectious diseases. All populations will be affected by climate change, but some are more vulnerable than others. Areas with weak health infrastructure, low socioeconomic status and elderly populations especially in developing countries will be the least able to cope with the hazardous effects of climate change.

**Keywords:** climate variables, temperature, humidity, health, mortality

## 1. Introduction

Over the past 50 years, burning fossil fuels have released sufficient quantities of greenhouse gases including carbon dioxide in the lower atmosphere to trap heat and affect global climate. Average world temperature has increased about 0.85°C in the past 130 years, and the last three decades has been warmer than decades before 1850 [1].

These climate changes have made sea levels rise, glaciers melt and precipitation patterns change. Climate change affects clean air, safe drinking water, sufficient food and secure shelter. Rising sea levels destroy homes, medical and other facilities. More than half of the world's population lives within 60 km of the coastal line and they may be forced to move [1] and leave their properties, which in turn can increase social displacement, mental disorders, unemployment and crime rates.

| Climate variable | Unit | Definition |
|---|---|---|
| Mean temperature | °C | The average temperature in the specified time frame |
| Maximum temperature | °C | The maximum temperature in the specified time frame |
| Minimum temperature | °C | The minimum temperature in the specified time frame |
| Apparent temperature (AT) | °C | AT combines temperature and humidity (and occasionally wind) into a single index for the assessment of human comfort in the warm season. It can be calculated in different ways [5] |
| Diurnal temperature range (DTR) | °C | The daily maximum temperature minus the daily minimum temperature within 1 day [4] |
| Wet-bulb temperature (Tw) | °C | The temperature at which air becomes saturated by evaporation at constant pressure [5] |
| Temperature humidity index (THI) | °F | Is calculated as $THI = 0.4(T + T_W) + 15$ T and Tw are in degrees Fahrenheit [5] |
| Discomfort index | – | See temperature humidity index (THI) [5] |
| Humidex | – | Canada's version of a comfort index is called the humidex (Hx) and is calculated as $Hx = T + 0.555(e - 10)$ where T is in degrees Celsius and e is in hPa [5] |
| Heat index (HI) | – | The HI is used to express summer comfort levels. It is based on a complex multiple regression equation that combines T and RH. It is highly correlated with both AT and THI [5] |
| Precipitation | millimeter | Water that falls to the ground as rain, snow, etc. |
| Heat/cold wave | | Extreme temperature ≥2 days [36] |
| Hot/cold wave duration | | The number of consecutive days on which the threshold was exceeded [34] |
| Heat/cold wave number | | The chronological number of a heat/cold wave in any given summer/winter [34] |

| Climate variable | Unit | Definition |
|---|---|---|
| Relative humidity (RH) | % | The amount of water vapor present in air expressed as a percentage of the amount needed for saturation at the same temperature [6] |
| Absolute humidity (AH) | gram per cubic meter | The water content of air at a given temperature [12] |
| Specific humidity | Unitless or expressed as parts per thousand | The ratio of the water vapor content of the mixture to the total air content on a mass basis [12] |
| Mixing ratio | Unitless or expressed as parts per thousand | Mixing ratio is the mass of moisture per mass of dry air |
| Dew point temperature (Td) | °C | When unsaturated air is cooled to saturation at constant pressure and without changing the air's moisture content that temperature is the dew point temperature. Dew point depends on the air's vapor pressure [5] |
| Dew point depression | °C | The difference between the temperature and dew point temperature at a certain height in the atmosphere. For a constant temperature, the smaller the difference, the more moisture there is, and the higher the relative humidity [5] |
| Vapor pressure | millibars or hPa (1 hPa = 100 Pascals = 1 mb) | The partial pressure exerted by water vapor in the atmosphere's gaseous mixture [5] |

**Table 1.** Climate variables used in epidemiological research.

Extreme weather events are becoming more intense and more frequent [1]. The overall health effects of a changing climate are likely to be severely negative. The number of weather-related natural disasters in the world has more than tripled since the 1960s. Every year, these disasters result in deaths, mainly in developing countries [1].

Variables rainfall patterns and floods can contaminate freshwater supplies, heighten the risk of waterborne diseases such as diarrhea and create breeding grounds for disease-carrying insects such as mosquitoes [1]. Climate change is also likely to decrease the production of agricultural products in many regions, and this will increase malnutrition and undernutrition [1].

Although thousands of years ago, Hippocrates suggested that climate has a wide range of effects on human health [2], serious investigation about the effect of climate on health has just happened in the recent 20 years and in light of robust scientific evidence about global climate change. Now, some authors describe "Climate" as a key determinant of human health [3].

Various effects of different climate variables including temperature, humidity, precipitation, wind direction and speed; on human health have been investigated. Some of the variables used in climate and health studies are summarized in **Table 1**.

Diurnal temperature range (DTR) is known as an important meteorological indicator. It shows weather stability and authors think it is associated with global climate change and urbanization [4].

The proper humidity variable in epidemiological and environmental health research should be selected based on the research questions. The most commonly used humidity variable is relative humidity [5]. However, researchers think this variable has limited use, should be used with caution and should be avoided in research about health conditions in which proximity to saturation is not relevant [5]. Relative humidity varies as a function of both the air water vapor content and air temperature; therefore, it is difficult to figure out which variable actually relates to the dependent variable. The complexities associated with the relative humidity variable may explain some of the contrary results of epidemiological studies about how humidity influences health outcomes [5].

Variables that include thermal components, such as relative humidity, dew point depression, and vapor pressure change severely by time of day and season [5]. Researchers should be cautious in incorporating the daily or seasonal average of these variables in statistical models as they do not represent the average moisture content of the air [5]. If the research question is related to the degree of saturation, then relative humidity or dew point depression are appropriate variables to use; but their dependence on daytime or changing weather situations should be taken care of [5].

Water vapor mass-based climate variables, such as specific humidity, absolute humidity, mixing ratio, dew point temperature and vapor pressure, are often highly correlated [5] and therefore should not be used simultaneously in statistical models. These variables are used when the air's actual moisture content is important [5].

Absolute (AH) and relative humidity (RH) are related to temperature. The hotter the air, the more water it can hold and therefore a much higher AH is achievable in warmer weather. However, the amount of water that cold air can carry is low, and therefore, the relative humidity can get higher and it feels more humid in cold weather [6]. In high RH, sweat does not easily evaporate because the air is pretty much saturated and temperature does not lose by sweating [6].

If human comfort in hot environments is the research question, then wet-bulb temperature or apparent temperature is the best variable to work with [5].

Davis et al. [5] made the following recommendations for choosing the right humidity variables in epidemiological research. First, the humidity variable should be chosen based on the research question and primary health consideration. For example, specific humidity can be used when the effect of atmospheric moisture content on disease is assessed. This variable might also be important for studies on microbial or fungal disease and pulmonary diseases. Apparent temperature can be used for studies about thermal stress such as heat shock or sudden cardiac death. Second, it is better to use several daily measurements (at least max and min values) rather than one measure for humidity variables; especially, when working in middle to high latitudes where there is large daily variation in humidity.

Researchers should also consider that the effects of climate variables can be modulated by other factors, such as social development, infrastructure, socioeconomic status and human adaptation [2].

In this chapter, we try to summarize the main health effects of climate variables reported in world studies.

## 2. Infectious diseases

Climate can determine the type of infectious diseases prevalent in different geographical areas, whereas weather can affect the time and the intensity of infectious disease outbreaks [3].

Several infectious diseases have been found related to climate variables. The more popular ones are listed below:

### 2.1. Malaria

Among different infectious diseases, the incidence of malaria, in particular, is generally thought to increase because of climate change and global warming [7]. Other vector-borne diseases may increase or decrease, but they currently make much less victims than malaria [7].

Diseases, such as malaria, which are transmitted by mosquito vectors, are sensitive to meteorological conditions. Excessive heat and cold kills mosquitoes. Malaria mosquitoes persist in a range between 17 and 33°C [8]. In this range, warmer temperatures increase mosquito reproduction and biting activity and the rate at which pathogens mature within them. For example, at 20°C, falciparum protozoa take 26 days to incubate, but at 25°C, they develop in 13 days. Also, Anopheles mosquitoes live only several weeks and warmer temperatures permit parasites to mature earlier, and the mosquitoes have more time to transfer the infection [3].

Temperature thresholds also limit the geographic range of mosquitoes. Transmission of falciparum malaria occurs in geographical areas where temperatures exceed 16°C [3].

Studies from Kerman, Iran, showed that the most effective meteorological factor on the incidence of malaria was temperature. As the mean, maximum and minimum of monthly temperature increased, the incidence rate raised significantly and models showed that a 1°C increase in maximum temperature in a given month was related to a 15 and 19% increase in malaria incidence on the same and subsequent month, respectively. Other studies from other world countries have also shown the effect of rising temperature in the incidence of malaria [9].

Dynamic models project that global warming will increase the transmission capacity of mosquitoes some 100-fold in temperate zones, and that the areas capable of sustaining transmission will grow and include more world populations [3]. The reports show that malaria has returned to South Korea, parts of southern Europe and the former Soviet Union. Malaria has also recolonized in the Indian Ocean, coastal province of South Africa, [3] many of these changes in the pattern of diseases are indicative of long-term warming and climate changes. Similarly, climate warming and the resulting change in the length of seasons in the East African highlands have led to an increased incidence of malaria [10].

Over the past century, intense precipitation (>5 cm over 24 h) has become more frequent, and warming of land surface has apparently intensified the monsoons that are strongly associated with mosquito and waterborne diseases in India and Bangladesh [3]. Several studies showed a positive association between increases in malaria and relative humidity, which is often positively correlated with precipitation [5].

In Gao et al.'s study, in Anhui Province, China, rainfall ($r_s = 0.48$) had the highest relation with malaria incidence. Malaria is a reemerging disease in this province, and rainfall is known as an important meteorological factor in the reemerging of this disease in the region. In this study, beside the effect of the same month's rainfall on malaria transmission, rainfall in the earlier 2 months also influenced malaria incidence [9]. Intense precipitation has also been reported to cause malaria outbreaks in Honduras (1998), Venezuela (1999) and Mozambique (2000) after hurricanes, torrential rains and cyclones in South America and southern Africa [3]. Climate change can allow diseases to invade immunologically naive populations with unprepared medical and health-care facilities [7].

However, very high rainfall can reduce mosquito populations by flushing larvae from their habitat in water swamps [11]. Researchers have also documented the association of malaria outbreaks with the El Niño Southern Oscillation(ENSO) cycle [11].

### 2.2. Yellow fever

Yellow fever is a climate-related viral disease that has a high rate of mortality and is carried by *Aedes aegypti*. Yellow fever is restricted by the 10°C winter isotherm and freezing kills Aedes eggs, larvae and adults [3].

### 2.3. Dengue fever

Studies suggest that there is a direct relationship between global warming and dengue fever [12]. Dengue fever is characterized by severe headaches and bone pain, and mortality occurs in case of hemorrhagic fever and shock syndrome. It is carried by *Aedes aegypti* and is restricted by the 10°C winter isotherm [3]. Climate change has helped dengue fever to spread into northern Australia and Argentina [3] as many of these changes in the pattern of diseases happened after long-term warming. Extreme weather and especially intense precipitation events after hurricanes have led to outbreaks of dengue fever in Honduras in 1998 and Venezuela in 1999 [3].

Changes in temperature and rainfall may also affect the distribution of disease vectors in dengue fever [2]. Researchers think that in the Asia-Pacific region, El Niño and La Niña events seem to affect the occurrence of dengue fever outbreaks [11].

### 2.4. Leishmaniasis

Studies have indicated that climate variability may influence changes in the vector geographical distribution as well as the density of the rodent reservoirs of leishmaniasis. In South America, climate variability based on ENSO revealed a significant effect on leishmaniasis. Also significant relationships were found between Mediterranean visceral leishmaniasis and climatic factors in some studies [5].

A study from Tunisia found that for relative humidity above 57.8% and lagged by 2 months, for each 1-unit increase in relative humidity, the disease incidence significantly increases by 5%. This study also showed seasonality during the same epidemiologic year and intervals between zoonotic cutaneous leishmaniasis (ZCL) epidemics ranging from 4 to 7 years. Mathematical models showed that ZCL incidence raises by 1.8% (95% CI: 0.0–3.6%) when there was 1-mm increase in the rainfall lagged by 12–14 months, and by 5.0% (95% CI: 0.8–9.4%) when there was a 1% increase in humidity from July to September in the same epidemiologic year. The researchers think that higher rainfall is expected to result in the increased density of plants that are food for *Psammomys obesus* (the reservoir rodent). Consequently, following an increase in the population of this rodent, the pool of *Leishmania major* transmissible from the rodents to blood-feeding female sand flies increases and can lead to a higher probability of transmission to humans over the next season [13].

### 2.5. Tick-related diseases

Warm winters have been demonstrated to facilitate northern migration of the ticks that carry tick-borne encephalitis and Lyme disease [3]. There is now evidence of vector species responding to recent climate change in Europe. For example, there has been latitudinal shifts in ticks, which carry tick-borne encephalitis in northern Europe [2]. Also tick-borne encephalitis has extended geographically in Sweden, and the tick vector of Lyme disease has spread in eastern Canada [11]. Tick-borne encephalitis in Sweden is likely related to warmer winters over the past two decades. The geographic range of ticks that transmit Lyme disease and viral encephalitis has extended to higher latitudes in Sweden and to higher altitudes in the Czech Republic [11].

Changes in climate that can affect the transmission of vector-borne infectious diseases include temperature, humidity, rainfall, soil moisture and sea level rise [1]. Research is ongoing to determine how these factors affect the risk of vector-borne diseases. Examples of vector-borne diseases likely to be sensitive to climate change has been shown in **Table 2**.

Crimean–Congo hemorrhagic fever (CCHF) is another tick-borne disease in Africa, Asia, Eastern Europe and the Middle East. It is a viral hemorrhagic fever transmitted mainly through tick bites and/or contact with blood and body fluids of patients (and/or infected animals). Studies from Iran showed that climate variables including mean temperature, accumulated rainfall and maximum relative humidity were significantly correlated with monthly incidence of CCHF. The number of cases in warmer summers was higher than the cooler ones, and also that the warmer the winters, the higher the number of cases [14].

| Vector | Diseases |
|---|---|
| Mosquitoes | Malaria, Filariasis, Dengue fever, Yellow fever, West Nile fever, Chikungunya fever |
| Sand flies | Leishmaniasis |
| Triatomines | Chagas disease |
| Ixodes ticks | Lyme disease, Tick-borne encephalitis |
| Hyalomma ticks | Crimean-Congo Hemorrhagic Fever (CCHF) |
| Tsetse flies | African trypanosomiasis |
| Black flies | Onchocerciasis |
| Snails | Schistosomiasis |

**Table 2.** Vectors and their related diseases that are likely to be sensitive to climate change [2, 14, 59].

The majority of cases of CCHF have been reported in Iran, Turkey and Bulgaria and correspond closely with the months that the temperature is between 30 and 40°C and maximum humidity is between 20 and 50% which is the favorite condition of the ticks. There are higher numbers of reported CCHF cases in warmer seasons and seasons with low rainfall [14].

### 2.6. Diarrhea and gastroenteritis

Changes in temperature and rainfall may affect the incidence of diarrheal diseases [2]. In tropical and subtropical regions with crowding and poverty, heavy rainfall and flooding may trigger outbreaks of diarrhea [11] by contaminating fresh water resources.

In Brisbane, Australia, there was a statistically significant positive relationship between diurnal temperature range (DTR) and diarrhea among children younger than five years. This effect was the greatest at one-day lag, with a 3% (95% CI: 2–5%) increase in emergency department admissions per 1°C increment of diurnal temperature range. The relative risk increased rapidly when DTRs were over 10°C [15]. Diarrheal diseases in Peru and Fiji have also accompanied short-term increases in temperature [11].

Ambient humidity has been reportedly associated with infectious enteritis. Studies in Japan, Taiwan and Peru showed negative relationships between relative humidity and infectious gastroenteritis [5]. Also similar relationships have been uncovered for rotavirus, another agent causing enteritis in Australia [5].

#### *2.6.1. Salmonella*

Salmonella bacteria proliferate more rapidly at higher temperatures and in animal gut and food [11]. Strong linear associations have been reported between temperature and notifications of salmonellosis in European countries and Australia [11].

Some recent studies have provided evidence about associations between weather events and the incidence of Salmonella. For example, studies have identified associations between average temperature and the number of reported cases of Salmonella infection. However, coastal communities may be more vulnerable because flooding events can contaminate their water supply with bacteria. A study from the United States observed a 4.1% increase in salmonellosis risk associated with a one-unit increase in extreme temperature events, and this increase in risk was more in coastal versus non-coastal areas (5.1% vs. 1.5%). Also they observed a 5.6% increase in salmonellosis associated with a one-unit increase in extreme precipitation events, and the effect was stronger in coastal areas (7.1% vs. 3.6%) [16].

Typhoid fever is a life-threatening illness caused by the bacterium *Salmonella* typhi. In February 2000, after torrential rains and a cyclone-inundated large parts of southern Africa and Mozambique, typhoid spread in the area [3].

#### *2.6.2. Cholera*

Studies have shown that cholera bacteria proliferate more rapidly at higher temperatures and in water [11]. Intense precipitation has been reported to cause outbreaks of cholera after hurricanes in Honduras in 1998, and after torrential rains and a cyclone in Mozambique in 2000 [3]. It is possible that increases in the rate of coastal outbreaks of cholera are also related to the warming of coastal waters and El Niño events [17].

A study from Iran showed that the incidence of cholera was significantly related to higher temperature and humidity and lower precipitation. Cholera epidemics are most likely to occur in hot seasons and in countries with more than one hot season, several cholera epidemics are likely each year. The significant relationship reported between the incidence of cholera and the lack of precipitation in Iran may be due to the fact that drought leads to the use of unsafe water [18].

### 2.7. Tuberculosis (TB)

In some countries, the highest incidence of diagnosed tuberculosis (TB) was in spring. Although the exact mechanism of this seasonal pattern is not well understood, there is a possibility that factors, such as temperature, humidity and sunlight, are related to TB incidence. Some researchers hypothesize that since winter is a cold season and people live in closed environments during winter; thus, transmission of TB happens in winter, and eventually, the symptoms and diagnosis happen in spring [19, 20].

A study from Kerman, Iran, reported that the incidence of TB increased in warm months, and for each one-unit C increase in temperature, the risk of TB increased 1.03 times. Also relative humidity with one-year lag had a reverse association with TB [21].

### 2.8. Hand, foot and mouth disease (HFMD)

HFMD is a common viral illness that usually affects children under 5 years old. Symptoms include fever, mouth sores and skin rashes. A study in China found that the commonly hot days positively affected the hand, foot and mouth disease (HFMD) burdens with the relative risk (RR) peaking at around 6 days of lag. The RR of HFMD in the Pearl River Delta Region was generally higher and persisted longer than that in the remaining developing areas [22].

### 2.9. Melioidosis

Melioidosis is an infectious disease that can infect humans and is caused by the bacterium *Burkholderia pseudomallei*. It is predominately a disease of tropical climates, especially Southeast Asia and northern Australia. The bacteria causing melioidosis are found in contaminated water and soil. It is spread to humans through direct contact with the contaminated source. Symptoms and signs of melioidosis can be mild, but severe manifestations such as bacteremia, organ abscesses and severe pneumonia can lead to death. Researcher found a significant correlation of melioidosis cases in Singapore with higher rainfall and, to a lesser degree, with higher humidity levels [23].

### 2.10. Other

Studies suggest a direct relationship between global warming and schistosomiasis [7]. Inter-annual and especially ENSO-related variations in climatic conditions in Australia have been reported to affect outbreaks of Ross River virus disease [11]. Climate also effects hantavirus pulmonary syndrome (HPS) and West Nile virus (WNV) [3].

## 3. Mortality

Normal human body temperature is maintained by the hypothalamus and is 36.1–37.8°C. When the environmental temperature exceeds the regulatory capacity of the hypothalamus, this can exert substantial stress on body organs [24].

Several world studies have shown that extreme temperatures can increase mortality. These graphs generally have the shape of a U, V or J and show an increase in mortality beyond a specific threshold temperature [24]. In most of these studies, a minimum mortality temperature ($T_{MM}$) or a comfort range, in which the least number of mortality per unit of time happens, has been reported.

In Greater Beirut, the $T_{MM}$ was 27.5°C and 1°C rise in temperature yielded a 12.3% increase (95% CI: 5.7–19.4%) and 1°C drop in temperature caused 2.9% increase (95% CI: 2–3.7%) in mortality [25]. The $T_{MM}$ in other world cities can be seen in **Table 3**.

Although temperature itself can effect mortality through physiological routes; low income, lack of air conditioning, poor access to transport, poor education, unhygienic microenvironments and older age have been recognized as risk factors which increase vulnerability to heat and cold [25]. Studies have shown that heat waves increase mortality more in vulnerable populations, such as elderly people, especially women, mentally ill people, children, those in thermally stressful occupations or people with preexisting illness [11].

It is very likely that climate change will lead to more frequent heat waves. Excess deaths were reported in England, Wales and France during the 2003 heat wave and caused a public health crisis. Much of the mortality from heat waves is due to cardiovascular, cerebrovascular and respiratory causes and happens more in the elderly [2].

Some researchers have mentioned a phenomenon called "urban heat island effect," which refers to urban centers with temperatures being somewhat higher than the surrounding suburban and rural areas [2]. Some inner urban environments have high thermal mass and low ventilation, which absorbs and retains heat and amplifies the rise in temperatures, especially overnight [11]. The impact of extreme heat on human health may also be exacerbated by increases in humidity [2].

Populations are likely to acclimatize to climate change through a range of behavioral, technological and physiological adaptations. However, infrastructural changes are likely to happen much slower, especially in developing countries [2].

The temperature–mortality relation varies greatly by latitude and climatic zone. People in hotter cities are more commonly affected by low temperatures, and people in colder cities are more affected by high temperatures. Other factors such as housing that may provide poor protection against cold or heat can cause higher excess winter mortality than expected [11]. However, cold also shows its deadly effect through infectious diseases such as influenza in elderly people or respiratory syncytial virus in infants [11].

| City | Latitude | $T_{MM}$ (°C) | Shape of curve |
|---|---|---|---|
| Salvador, Brazil [26] | 12.97 S | 23 | J |
| Bangkok, Thailand [26] | 13.75 N | 29 | J |
| Chiang Mai, Thailand [26] | 18.79 N | 19–28 | U |
| Mexico City, Mexico [26] | 19.43 N | 15–18 | Wide U |
| Taishan, China [8] | 22.25 N | 25.7 | J |
| Zhuhai, China [60] | 22.27 N | 25.9 | J |
| Shenzhen, China [29] | 22.55 N | 33 | J |
| Guangzhou, China [60] | 23.13 N | 26 | J |
| Sao Paulo, Brazil [26] | 23.55 S | 21–23 | Wide U |
| Taipei, Taiwan [27] | 25.03 N | 25.2–31.5 | Asymmetric V |
| Nanxiong, China [8] | 25.11 N | 24 | J |
| Monterrey, Mexico [26] | 25.66 N | 17–31 | Asymmetric U |
| Miami, USA [25] | 25.77 N | 27.2 | |
| Tampa, Florida, USA [61] | 27.96 N | 27.1 | |
| New Delhi, India [26] | 28.61 N | 19–29 | J |
| Chongqing, China [29] | 29.55 N | 34 | J |
| Shiraz, Iran [41] | 29.61 N | 20–25 | J |
| Kerman, Iran [40] | 30.28 N | 21.1–25.1 | Wide J |
| Jacksonville, Florida, USA [61] | 30.33 N | 24.8 | |
| Shanghai, China [38] | 31.20 N | 28 | Reversed J |
| Nanjing, China [29] | 32.05 N | 35 | J |

| City | Latitude | $T_{MM}$ (°C) | Shape of curve |
|---|---|---|---|
| Santiago, Chile [26] | 33.45 S | 16 | Wide U |
| Tehran, Iran [40] | 33.69 N | 28.5 | |
| Beirut, Lebanon [25] | 33.88 N | 27.5 | Wide asymmetric V |
| Cape Town, South Africa [26] | 33. 92 S | 17 | Wide U |
| Tokyo, Japan [27] | 35.68 N | 29.4–30.8 | V |
| Seoul, South Korea [27] | 37.56 N | 30.1–33.5 | Asymmetric V |
| Seoul, South Korea [24] | | 29.5 | J |
| Athens, Greece [35] | 37.58 N | 22.7–25.7 | |
| Baltimore, USA [25] | 39.28 N | 21.4 | |
| Valencia, Spain [25] | 39.46 N | 15 (winter), 24 (summer) | |
| Beijing, China [27] | 39.91 N | 31.3–32.3 | J |
| Beijing, China [38] | | 25 | U |
| Castile-La Mancha, Spain [34] | 40.10 N | 37 | |
| Boston, USA [25] | 42.36 N | 21 | |
| Sofia, Bulgaria [26] | 42.70 N | 16 | J |
| Christchurch, New Zealand [40] | 43.53 S | 20.5 | |
| Bucharest, Romania [26] | 44.42 N | 22 | Wide U |
| Harbin, China [29] | 45.75 N | 29 | J |
| Ljubljana, Slovenia [26] | 46.05 N | 17 | J |
| Kings County, Washington, USA [30] | 47.47 N | seems like 22.1 | J |
| London, UK [35] | 51.50 N | 19.3–22.3 | |
| Holland [25] | 52.31 N | 16.5 | |
| North Finland [35] | 67 N | 14.3–17.3 | |

**Table 3.** The reported minimum mortality temperature, latitude and the shape of the mortality curve, in different world cities sorted based on latitude.

McMichael et al. estimated the temperature threshold below which cold-related mortality begins to increase, to range from 15 to 29°C, and the threshold for heat-related deaths to range from 16 to 31°C in different world cities. These researchers found heat thresholds were generally higher in cities with warmer climates, but cold thresholds were unrelated to climate [26]. Other researchers have reported lower latitude cities to have higher threshold temperatures [27]. The reported minimum mortality temperature for some world cities, sorted by latitude, has been shown in **Table 3**.

http://dx.doi.org/10.5772/64955

A meta-analysis showed that the effect of cold on mortality were delayed and lasted at least 10 days, whereas heat effects appeared quickly and lasted usually only 3–4 days. Interestingly, despite widely ranging climates, the $T_{MM}$ were close to the 75th percentile of temperature in all 12 countries/regions studied in the meta-analysis, suggesting that people have probably adapted to their local climates [28]. This finding is consistent with $T_{MM}$ in communities with colder climates being lower than in communities with warmer climates [28].

In China, associations between daily maximum temperature and daily mortality from all-causes were observed in different cities, with increases in 3.2–5.5%, with each 1°C increase in the daily maximum temperature over the threshold. Also a stronger temperature-associated mortality was detected in females and adults over 30 years [29]. Isaksen et al. [30] in King County, Washington, showed that heat, expressed as humidex, is associated with increased mortality and that the risk increases with heat's intensity.

In China, researchers also observed statistically significant associations of DTR with total, cardiovascular and respiratory mortality in most cities in the full year and in cool seasons. However, few significant results were found in warm seasons. The researchers think that wide DTR might be a source of additional stress on the cardiorespiratory systems in low temperatures, and this stress might have more detrimental effects in older people and those with underlying cardiovascular disease [31]. Increase in mortality with increase in DTR was also seen in other Chinese cities and a multicity study in Korea. Researchers have suggested that the consistency in the literature shows the association of DTR with mortality are not likely to be substantially changed by geography, climate, population, publication bias or model specifications [31].

A study from a coastal city in India showed a clear effect of ambient heat in the increase in all-cause mortality and suggested that heat index has a stronger effect than maximum temperature on mortality. This study also showed an inverse relationship between mean mortality and relative humidity [32].

In a high plateau area in southwest China, risk assessments showed a strong increase in mortality starting at a DTR of approximately 16°C. The risk of mortality with extreme high DTR was greater for males and aged under 75 years. Researchers suggested that DTR of 16°C may be a good cutoff point for epidemiological mortality studies [4].

Researchers in Australia found that winters that were colder or drier had significantly increased death risks in most cities, whereas warmer or more humid summers did not increase the risk of death. The strongest increase in deaths for a colder winter was in Brisbane, the city with the warmest climate and the mildest winter. This again shows that warmer cities are more vulnerable to cold. Also, studies have showed that drier winters are associated with more influenza outbreaks [33].

Linares et al. [34] in Spain found that the variable, heat wave duration, was of major importance in mortality. The significant lags between temperature and mortality during heat waves range from 0 to 6 days. But for cold waves, this impact was extended to 12-day lag for respiratory deaths [34].

The variable, relative humidity, is usually present in mortality models with a negative sign showing inverse association and is thought to reduce the effect of heat and cold on mortality. Some researchers think that it is preferable to address temperature and relative humidity separately in epidemiological studies, especially when it comes to defining heat and cold waves and estimating their effect [34]. However, in Shanghai, Philadelphia and Sydney extreme maritime tropical (warm and moist) air mass was associated with high mortality, indicating that extreme humidity may be as dangerous as extreme temperature [5].

Researchers have reported that the $T_{MM}$ in different world cities correlates with the latitude. Others have mentioned that $T_{MM}$ varies with temperature of the place and humans adapt to local climatic conditions. In France, 80% of $T_{MM}$ variance was accounted for by the mean summer temperatures (MST), and $T_{MM}$ was highest in the southern parts of France [35].

In Australia, pooled data show that the relative risk of mortality started to increase around the 95th percentile of temperature, increased sharply at the 97th percentile and rose alarmingly at the 99th percentile. These researchers think that tiered health risk-based metrics should be performed to define a heat wave [36].

A study about infant mortality in California showed that the excess mortality risk was 4.4% per 5.6°C increase for average of same day and a previous 3-day apparent temperature. The associations for apparent temperature were highest for black infants. This study suggested that infants were also a vulnerable subgroup to heat exposure [37].

### 3.1. Cardiovascular disease and mortality

In China, strong associations between daily maximum temperature and daily mortality from cardiovascular causes were observed in different geographical cities, with increases in 4.6–7.5% with each 1°C increase in the daily maximum temperature over the threshold [29].

In Beijing, people with hypertensive disease were susceptible to both extremely low and high temperatures, and in Shanghai, people with ischemic heart disease showed greater susceptibility to extremely cold days [38]. Some studies have documented an association between mean temperature and humidity variations, and the number of visits to the emergency departments for atrial fibrillation [10].

In East Asia, heat waves had the strongest effects on cardiovascular deaths, which was (8.8, 95% CI: 5.5–12.2) [27]. In Washington State, statistically significant results were found for circulatory (9%) and cerebrovascular (40%) deaths and heat in all ages [30] and stratifying by age, and statistically significant increases in mortality risk on hot days were found for the 65–84 age group, in cerebrovascular (37%), and in the 85+ age group, in circulatory (18%), cardiovascular (17%), and cerebrovascular (53%) mortality [30].

In China per capita years of education (as an indicator of economic status), percentage of population over 65 years and percentage of women had direct impact on cold-related cardiovascular mortality in populations. Also number of hospital beds (as an indicator of the availability of medical resources), percentage of population engaged in industrial occupations, and percentage of women showed direct impact on heat-related cardiovascular mortality [39], which confirms that socioeconomic factors can alter the effect of climate variables on cardiovascular mortality.

Gender also shows a different impact at low and high temperatures. Men tend to have a higher risk at low temperatures, whereas women tend to have higher risk at high temperatures [39].

A study from Kerman, Iran, showed increases in daily mortality from cardiovascular diseases as temperature decreased. Also significant correlations were observed between cardiovascular mortality and temperature, and the maximum correlations for cardiovascular deaths were on lag 0–lag 3. For each 1°C decrease in temperature, cardiovascular deaths showed a 0.6% increase [40], but no increase in cardiovascular mortality was detected with increased temperature, which is probably related to acclimatization. In Shiraz, Iran, the minimum number of cardiovascular deaths happened at 20°C. Drops in mean monthly temperature were significantly associated with increased 18- to 60-year-old cardiovascular deaths that happened one month later [41].

### 3.2. Respiratory disease and mortality

There is epidemiological evidence that shows influenza-related morbidity and mortality peaks 2–3 weeks after falls in AH. Also, in vitro experiments have shown improved survival of the influenza virus at lower AH levels [6]. Extremes can be hazardous for health in many other indirect ways as well. Prolonged droughts fuel bush fires that release hazardous respiratory pollutants [3].

In Korea, above a threshold temperature of 29.5°C, a rise in temperature of 1°C resulted in an increase in death from respiratory conditions (RR 1.02; 95% CI: 1–1.04). There was also an increased risk of death from asthma (RR 1.05, 95% CI: 1.01–1.11) [24].

In Hong Kong, cold temperature and rainfall was associated with most influenza epidemics; but, relative humidity and absolute humidity did not show much contribution to epidemics [42]. This effect may be due to prolonged survival of viral particles under colder conditions or enhance crowding and indoor activities that would increase contact, aerosol and droplet transmission [42].

Some studies have shown that rainfall could be a predictor to forecast influenza infection for subtropical and tropical regions, but not in all temperate regions. One plausible mechanism is that rainfall could increase indoor activities, and therefore influence the number of contacts and the risk of exposure to infected individuals [42].

A study from Turkey reported that some meteorological parameters such as wind direction, air temperature and atmospheric pressure were related to the incidence of pulmonary embolism. But no relation was found between unprovoked pulmonary embolism(PE) cases' monthly distribution and pressure, humidity or temperature. However, there was a statistically significant positive correlation between provoked PE cases and air temperature [43]. The relation between PE and hot temperature may be related to dehydration or people traveling in cars for longer distances [43]. In a study about temperature and infant mortality, white infants had an elevated risk for deaths from respiratory causes [37].

Climate-related events including heat waves and extreme meteorological events can increase the frequency of acute cardiorespiratory events due to higher concentrations of ground level ozone, changes in particle pollution, altered spatial and temporal distribution of allergens (pollens, molds, and mites), and some infectious disease vectors. These events will not only aggravate the condition of those with current respiratory disease and asthma but also increase the incidence and prevalence of allergic respiratory conditions [44].

Weather can affect asthma directly, by acting on airways, or indirectly, by influencing airborne allergens and pollutant levels. Cold air temperature can aggravate asthmatic symptoms [44]. There is evidence that, during pollen season, thunderstorms can be associated with asthma outbreaks or acute respiratory disease outbreaks [44, 45].

Some studies have reported higher barometric pressure, more hours of sunshine and lower humidity in winter to be associated with an increase in chronic obstructive pulmonary disease (COPD) exacerbations, implying that warm and dry high pressure systems were associated with COPD anomalies. Studies from Trinidad showed that in warm, wet climates incidence of asthma increased with higher relative humidity in the wet season. Conversely, a study from Japan demonstrated an association between low relative humidity and hospital admissions for pediatric asthma. The other indirect effects of humidity in respiratory disease include its role in promoting the increasing mold and mites [5]

In many world countries, low humidity levels were found to precede the onset of increased winter time influenza-related mortality by several weeks. Low humidity probably impacts on virus stability and viability, host susceptibility and human behavior [5].

A study from Australia about pediatric emergency department visits showed that high temperatures had a significant impact on pediatric diseases, including chronic lower respiratory diseases. Low temperatures were also significantly associated with respiratory diseases [46].

A study from Kerman, Iran, showed increases in daily mortality from respiratory diseases as temperature decreased. This relation reached a maximum after a 26-day lag. In this study, for each 1°C decrease in temperature, respiratory deaths showed an average of 2.5% increase [40]. In Shiraz, Iran, the minimum number of respiratory deaths happened in 25°C. Mean monthly temperature was inversely and significantly associated with total and female respiratory deaths on the same month and with total, male and female respiratory deaths that happened one month later [41].

## 4. Premature delivery

In the United States, ambient temperature was significantly associated with preterm birth, and regardless of their maternal demographic characteristics or baby gender, each 5.6°C (10° F) increase in weekly average apparent temperature (with lags up to one week), caused an 8.6% increase (95% confidence interval: 6.0, 11.3) in preterm delivery. Preterm delivery has many etiologies, but one possible explanation for its relation with heat is increased dehydration with heat exposure, which may decrease uterine blood flow and increase pituitary oxytocin to induce labor [47].

In Spain, when maximum apparent temperature exceeded the 90th percentile, the risk of preterm birth increased up to 20% after 2 days, and when minimum temperature rose to the 90th percentile, it increased by 5% after a week [48]. Exposure to moderately high temperatures during late pregnancy might be associated with an increase in risk of preterm birth [49].

In Rome and Barcelona, increase in maximum apparent temperature (MAT), especially in the second half of the second trimester, increased the risk of preterm and particularly early preterm births [50].

## 5. Diabetes, endocrine and metabolic diseases

Some researchers have suggested that climate change may be related to increase in type 2 diabetes [51].

Studies have reported significant associations between increases in daily endocrine and metabolic diseases mortality with increase in the daily maximum temperature above the threshold. Mortalities for diabetes were also significantly associated with temperature. The increased mortality for every 1°C increase in the daily maximum temperature over the threshold for endocrine and metabolic outcomes, and particularly diabetes, was 12.5–31.9% and 14.7–29.2% [29]. Statistically significant increases in post-heat exposure diabetes-related mortality in the 45–64 age group in the United States suggests that underlying health status may contribute to these risks [30].

In a study about pediatric emergency department visits in Australia, high temperatures had a significant impact on endocrine and metabolic pediatric diseases, whereas low temperatures were also significantly associated with endocrine, nutritional and metabolic diseases [46].

Although congenital hypothyroidism was reported to have a seasonal pattern in some parts of the world, a recent study did not find a significant pattern [52].

## 6. Mental diseases

Extremes of temperature and rainfall, such as heat waves, floods and drought, have both direct immediate effects such as mortality, and indirect longer term effects. For example, populations that have survived severe floods and drought may suffer from sustained increases in common mental disorders [2].

In a study about the effects of extreme heat in mortality in the United States, statistically significant results were found for mortality related to mental disorders which showed a 43% increased risk [30].

## 7. Injuries and trauma deaths

Extreme events such as floods can cause injuries, deaths and other sequelae [11]. A study from Iran showed that the overall mortality caused by trauma was higher in the warm season, and the highest significant correlation between unintentional trauma deaths and temperature was seen in ages over 60 years ($r = 0.301$). Also, an inverse significant correlation was observed between the unintentional trauma deaths and humidity and was again highest in the over 60-year age group ($r = -0.336$). The authors think these results may be attributed to increase in activity or travelling in warm seasons and increased risk of unintentional injuries, such as traffic accidents, falls, drowning and heat exhaustion. Also, older people tolerate hot environments less than others [53].

Some studies have shown a peak in suicide rates during the spring season, and attributed it to increased temperatures; and others showed no relation between suicide and temperature. Some authors have regarded pollen as an important seasonal aeroallergen that may act as trigger for suicide [53]. However, suicide can be related to many other important socioeconomic factors that have to been considered in these studies. Some researchers have commented that suicide is a complex, phenomenon driven by not only biological factors but also interactions between individuals and their environments, and weather variables may only increase the risk [53].

In a study about the effects of extreme heat on mortality in the United States, statistically significant results were found for accidental deaths [30].

## 8. Parkinson's disease

In Spain, a study about the effect of heat waves on Parkinson's disease (PD) mortality and morbidity showed that at a maximum daily temperature of 30°C, PD-related admissions were at a minimum. But starting at a temperature of 34°C, the number of admissions increases with temperature. Researchers concluded that suffering from PD is a risk factor for excess morbidity and mortality associated with high temperatures [54].

## 9. Multiple sclerosis

A study from Kerman, Iran, showed that the highest number of hospital admissions for multiple sclerosis happened in spring and winter, and this seasonal pattern was more pronounced in women. Researchers think that the seasonal-related hospital admissions are probably related to climate variables or seasonal infectious diseases [55].

Other researchers have found that the disease prevalence is lower in warmer climates, which enjoy more sunshine; also as latitude and distance from the equator increases, the prevalence of MS increases as well. Others have shown that the prevalence is less in people with sun burns or those who have high levels of vitamin D [55]. However, more research about the role of climate variables in the incidence of MS is needed as some studies have not shown these relations.

## 10. Allergic diseases

Climate change may change the timing and duration of the pollen and spore seasons and the geographic scope of these aeroallergens, affecting allergic disorders such as hay fever and asthma [11].

Meteorological events can alter the onset, spatial and temporal distribution and the duration and intensity of allergens such as pollens, molds and mites. Therefore, the onset, duration and intensity of the pollen may also vary from year to year. Weather variables are among the main factors affecting phenology and pollen production by plants [44].

Different climate variables influence the daily fluctuations in the amount of pollen. The more important variables are temperature, rainfall and duration of sunshine. At least 10 weather elements are thought to affect the concentration of pollen, which are temperature, rainfall, average wind speed, relative humidity, maximum temperature, minimum temperature, temperature range, continued rainfall hours, accumulated sunshine hours and accumulated mean temperature [44].

When conditions are good for pollination (ripe anthers, low humidity and warm temperature), anthers open and release pollen. If the favorable weather conditions arrive early, ripe anthers will release less ripe pollen with less allergen. Otherwise, if weather is non-favorable and anthers do not open until later, riper pollen with more allergens are released [44].

These events can increase the incidence and prevalence of allergic-related conditions [44]. Thunderstorms can also induce attacks of severe asthma and are a common cause of epidemics of asthma attacks requiring Emergency Department visits [44]. Thunderstorm-related asthma has happened in England, Canada, Mexico, Australia, Italy [44] and Iran [45].

## 11. Renal problems

A US study on older adults showed that risks of hospitalization for fluid and electrolyte disorders, renal failure, urinary tract infection, septicemia and heat stroke were statistically significantly higher on heat wave days. For fluid and electrolyte disorders and heat stroke, the risk of hospitalization increased during more intense and longer lasting heat wave periods. Risks were generally highest on the heat wave day but remained elevated for up to five subsequent days [56].

Another US study showed statistically significant increased mortality risk associated with extreme heat from nephritis and nephrotic syndromes [30]. In South Korea, a significant heat-associated increase in the RR of mortality from genitourinary conditions was observed. This shows that patients with preexisting chronic conditions may be more susceptible to high ambient temperatures [24].

Kidney stones have also been inversely linked to relative humidity in a few studies [5]. A study showed that extraterrestrial radiation, isothermality, annual mean temperature (AMT) and precipitation seasonality were significant predictors of urolithiasis prevalence in Iran, and urolithiasis is more prevalent in the south of Iran, which has a warmer climate. High temperatures can result in increased urinary concentration and low urinary volume due to excessive sweating, which can increase in the concentration of relatively insoluble salts that turn into stones. Some authors have indicated a strong relationship between annual mean temperature (AMT) and stone prevalence and predicted that cases of nephrolithiasis will increase due to the global warming [57].

## 12. Environmental toxins

Studies about climate change have found out that ocean warming around the Faroe Islands has facilitated the methylation of mercury and its subsequent uptake by fish. Researchers think that methyl mercury concentrations in fish will increase by 3–5% for a 1°C rise in water temperature. Eating methyl mercury-contaminated fish has harmful effects for humans and also impairs fetal/infant neurocognitive development [11, 58].

The severe effects of climate change and global warming on human populations suggest that actions should be taken to reduce its burden on human populations. All populations will be affected by climate change, but some are more vulnerable than others. Countries with weak health infrastructure, mostly in developing countries will be the least able to cope with climate change [1].

Many policies have the potential to reduce greenhouse gas emissions. For example, promoting clean energy, the safe use of public transportation and physical activity can reduce carbon emissions [1]. Carbon taxation has also been implemented in some developed countries.

## Author details

Narges Khanjani[1,2*]

Address all correspondence to: khanjani@umd.edu; n_khanjani@kmu.ac.ir

1 Maryland Institute of Applied Environmental Health (MIAEH), School of Public Health, University of Maryland, College Park, MD, USA

2 Environmental Health Engineering Research Center, School of Public Health, Kerman University of Medical Sciences, Kerman, Iran

## References

[1] World Health Organization, Climate Change and Health, Geneva. September 2015 [February 14, 2016]; Available from: http://www.who.int/mediacentre/factsheets/fs266/en/.

[2] Haines A, Kovatsa RS, Campbell-Lendrumb D, Corvalan C. Climate change and human health: Impacts, vulnerability and public health. Public Health 2006;120:585–96. doi: 10.1016/j.puhe.2006.01.002

[3] Epstein PR. Climate change and emerging infectious diseases. Microbes and Infection. 2001;3:747–54.

[4] Ding Z, Guo P, Xie F, Chu H, Li K, Pu J, et al. Impact of diurnal temperature range on mortality in a high plateau area in southwest China: A time series analysis. The Science of the Total Environment. 2015;526:358–65. doi:10.1016/j.scitotenv.2015.05.012

[5] Davis RE, McGregor GR, Enfield KB. Humidity: A review and primer on atmospheric moisture and human health. Environmental Research 2016;144(Pt A):106–16. doi: 10.1016/j.envres.2015.10.014

[6] Metz JA, Finn A. Influenza and humidity – Why a bit more damp may be good for you! The Journal of infection. 2015;71(Suppl 1):S54–8. doi:10.1016/j.jinf.2015.04.013

[7] Bosello F, Roson R, Tol R. Economy-wide estimates of the implications of climate change: Human health. Ecological Economics 2006;58:579–91. doi:10.1016/j.ecolecon.2005.07.032

[8] Beck-Johnson L, Nelson W, Paaijmans K, Read A, Thomas M, Bjørnstad O. The Effect of temperature on anopheles mosquito population dynamics and the potential for malaria transmission. PLoS One 2013;8(11):e79276. doi:10.1371/journal.pone.0079276

[9] Mohammadkhani M, Khanjani N, Bakhtiari B, Sheikhzadeh K. The relation between climatic factors and malaria incidence in Kerman, South East of Iran. Parasite Epidemiology and Control 2016. doi:10.1016/j.parepi.2016.06.001.

[10] Franchini M, Mannucci PM. Impact on human health of climate changes. European Journal of Internal Medicine 2015;26:1–5. doi:10.1016/j.ejim.2014.12.008

[11] McMichael AJ, Woodruff RE, Hales S. Climate change and human health: present and future risks. Lancet 2006;367:859–69.

[12] Wikipedia. Humidity. Beck Johnson 2013 [updated Jan 2014; cited 2016 Feb 10, 2016]; Available from: https://en.wikipedia.org/wiki/Humidity.

[13] Toumi A, Chlif S, Bettaieb J, Ben Alaya N, Boukthir A, Ahmadi ZE, et al. Temporal dynamics and impact of climate factors on the incidence of zoonotic cutaneous leishmaniasis in central Tunisia. PLoS Neglected Tropical Diseases. 2012;6(5):e1633. doi: 10.1371/journal.pntd.0001633

[14] Ansari H, Shahbaz B, Izadi S, Zeinali M, Tabatabaee SM, Mahmoodi M, et al. Crimean-Congo hemorrhagic fever and its relationship with climate factors in southeast Iran: a 13-year experience. The Journal of Infection in Developing Countries 2014;8(6):749–57. doi:10.3855/jidc.4020

[15] Xu Z, Huang C, Turner L, Su H, Qiao Z, al. e. Is diurnal temperature range a risk factor for childhood diarrhea? PLoS One. 2013;8(5):e64713. doi:10.1371/journal.pone.0064713

[16] Jiang C, Shaw KS, Upperman CR, Blythe D, Mitchell C, Murtugudde R, et al. Climate change, extreme events and increased risk of salmonellosis in Maryland, USA: Evidence for coastal vulnerability. Environment International 2015;83:58–62. doi:10.1016/j.envint.2015.06.006

[17] McMichael AJ. Globalization, climate change, and human health. The New England Journal of Medicine 2013;368(14):1335–43. doi:10.1056/NEJMra1109341

[18] Pezeshki Z, Tafazzoli-Shadpour M, Mansourian A, Eshrati B, Omidi E, Nejadqoli I. Model of cholera dissemination using geographic information systems and fuzzy clustering means: Case study, Chabahar, Iran. Public Health 2012;126:881e7. doi: 10.1016/j.puhe.2012.07.002

[19] Moosazadeh M, Khanjani N, Bahrampour A, Nasehi M. Does tuberculosis have a seasonal pattern among migrant population entering Iran? International Journal of Health Policy and Management 2014;2:181–85. doi:10.15171/ijhpm.2014.43

[20] Moosazadeh M, Khanjani N, Bahrampour A. Seasonality and temporal variations of tuberculosis in the north of Iran. Tanaffos. 2013;12(4):35–41.

[21] Fallah Ghoshghalb S, Khanjani N, Samareh Fekri M, editors. The effect of climate on tuberculosis. In: The 15th National Conference on Environmental Health; 2012 Oct 30–Nov 1; Rasht, Iran.

[22] Zhang W, Dua Z, Zhang D, Yu S, Hao Y. Quantifying the adverse effect of excessive heat on children: An elevated risk of hand, foot and mouth disease in hot days. Science of the Total Environment 2016;541:194–99.

[23] Liu X, Pang L, Sim SH, Goh KT, Ravikumar S, Win MS, et al. Association of melioidosis incidence with rainfall and humidity, Singapore, 2003–2012. Emerging Infectious Diseases. 2015;21(1):159–62. doi:10.3201/eid2101.140042

[24] Kim CT, Lim YH, Woodward A, Kim H. Heat-attributable deaths between 1992 and 2009 in Seoul, South Korea. PLoS One. 2015;10(2):e0118577. doi:10.1371/journal.pone.0118577

[25] El-Zeina A, Tewtel-Salemb M, Nehme G. A time-series analysis of mortality and air temperature in Greater Beirut. Science of the Total Environment 2004;330:71–80.

[26] McMichael AJ, Wilkinson P, Kovats RS, Pattenden S, Hajat S, Armstrong B, et al. International study of temperature, heat and urban mortality: the 'ISOTHURM' project. International Journal of Epidemiology 2008;37(5):1121–31. doi:10.1093/ije/dyn086

[27] Chung JY, Honda Y, Hong YC, Pan XC, Guo YL, Kim H. Ambient temperature and mortality: an international study in four capital cities of East Asia. The Science of the Total Environment 2009;408(2):390–96. doi:10.1016/j.scitotenv.2009.09.009

[28] Guo Y, Gasparrini A, Armstrong B, Li S, Tawatsupa B, Tobias A, et al. Global variation in the effects of ambient temperature on mortality. A Systematic Evaluation. Epidemiology 2014;25:781–89. doi:10.1097/EDE.0000000000000165

[29] Li Y, Cheng Y, Cui G, Peng C, Xu Y, Wang Y, et al. Association between high temperature and mortality in metropolitan areas of four cities in various climatic zones in China: A time-series study. Environmental Health 2014;13:65. doi:10.1186/1476-069X-13-65

[30] Isaksen TB, Fenske RA, Hom EK, Ren Y, Lyons H, Yost MG. Increased mortality associated with extreme-heat exposure in King County, Washington, 1980–2010. International Journal of Biometeorology 2016;60:85–98. doi:10.1007/s00484-015-1007-9

[31] Zhou X, Zhao A, Meng X, Chen R, Kuang X, Duan X, et al. Acute effects of diurnal temperature range on mortality in 8 Chinese cities. Science of the Total Environment 2014;493:92–7.

[32] Desai VK, Wagle S, Rathi SK, Patel U, Desai HS, Khatri K. Effect of ambient heat on all-cause mortality in the coastal city of Surat, India. Current Science 2015;109(9):1680.

[33] Huang C, Chu C, Wang X, Barnett AG. Unusually cold and dry winters increase mortality in Australia. Environmental Research 2015;136:1–7. doi:10.1016/j.envres.2014.08.046

[34] Linares C, Diaz J, Tobias A, Carmona R, Miron IJ. Impact of heat and cold waves on circulatory-cause and respiratory-cause mortality in Spain: 1975–2008. Stochastic Environmental Research and Risk Assessment 2015;29:2037–46. doi:10.1007/s00477-014-0976-2

[35] Todd N, Valleron AJ. Space-time covariation of mortality with temperature: A systematic study of deaths in France, 1968–2009. Environmental Health Perspectives 2015;123(7):659–64. doi:10.1289/ehp.1307771

[36] Tong S, FitzGerald G, Wang XY, Aitken P, Tippett V, Chen D, et al. Exploration of the health risk-based definition for heatwave: A multi-city study. Environmental Research 2015;142:696–702. doi:10.1016/j.envres.2015.09.009

[37] Basu R, Pearson D, Sie L, Broadwina R. A case-crossover study of temperature and infant mortality in California. Paediatric and Perinatal Epidemiology 2015;29:407–15. doi:10.1111/ppe.12204

[38] Wang X, Li G, Liu L, Westerdahl D, Jin X, Pan X. Effects of extreme temperatures on cause-specific cardiovascular mortality in China. International Journal of Environmental Research and Public Health 2015;12(12):16136–56. doi:10.3390/ijerph121215042

[39] Yang X, Li L, Wang J, Huang J, Lu S. Cardiovascular mortality associated with low and high temperatures: determinants of inter-region vulnerability in China. International Journal of Environmental Research and Public Health 2015;12(6):5918–33. doi: 10.3390/ijerph120605918

[40] Khanjani N, Bahrampour A. Temperature and cardiovascular and respiratory mortality in desert climate. a case study of Kerman, Iran. Iranian Journal of Environmental Health Science and Engineering. 2013;10(11). doi:10.1186/1735-2746-10-11

[41] Khanjani N, Dadbakhsh M, Bahrampour A. The effect of temperature on respiratory and cardiovascular mortality in Shiraz, Iran. In: The 26th ISEE Conference; 2014 August 24–28; Seattle, Washington, USA.

[42] Chong KC, Goggins W, Zee BC, Wang MH. Identifying meteorological drivers for the seasonal variations of influenza infections in a subtropical city—Hong Kong. International Journal of Environmental Research and Public Health 2015;12(2):1560–76. doi: 10.3390/ijerph120201560

[43] Anar C, İnal T, Erol S, Polat G, Ünsal I, Ediboğlu O, et al. Are meteorological parameters a risk factor for pulmonary embolism? A retrospective analysis of 530 patients. Balkan Medical Journal 2015;32:279–84.

[44] D'Amato G, Holgate ST, Pawankar R, Ledford DK, Cecchi L, Al-Ahmad M, et al. Meteorological conditions, climate change, new emerging factors, and asthma and related allergic disorders. A statement of the World Allergy Organization. The World Allergy Organization Journal 2015;8(1):25. doi:10.1186/s40413-015-0073-0

[45] Idani E, Dastoorpoor M, Goudarzi G, Khanjani N. Severe Outbreaks of respiratory syndromes following autumn rainfall in Khuzestan, Iran. Archives of Iranian Medicine (in press)

[46] Xu Z, Hu W, Su H, Turner LR, Ye X, Wang J, et al. Extreme temperatures and paediatric emergency department admissions. Journal of Epidemiology and Community Health 2014;68(4):304–11. doi:10.1136/jech-2013-202725

[47] Basu R, Malig B, Ostro B. High ambient temperature and the risk of preterm delivery. American Journal of Epidemiology. 2010;172(10):1108–17. doi:10.1093/aje/kwq170

[48] Vicedo-Cabrera AM, Iñíguez C, Barona C, Ballester F. Exposure to elevated temperatures and risk of preterm birth in Valencia, Spain. Environmental Research 2014;134:210–7.

[49] Vicedo-Cabrera AM, Olsson D, Forsberg B. Exposure to seasonal temperatures during the last month of gestation and the risk of preterm birth in Stockholm. International

Journal of Environmental Research and Public Health 2015;12:3962–78. doi:10.3390/ijerph120403962

[50] Schifano P, Asta F, Dadvand P, Davoli M, Basagana X, Michelozzi P. Heat and air pollution exposure as triggers of delivery: A survival analysis of population-based pregnancy cohorts in Rome and Barcelona. Environment International 2016;88:153–9. doi:10.1016/j.envint.2015.12.013

[51] Shubair MM, Haider M, Bassa M. Climate change and type 2 diabetes. Journal of Endocrinology and Diabetes Mellitus. 2013;1:22–6.

[52] Khanjani N, Ahmadzadeh A, Bakhtiari B, Madadizadeh F. The role of season and climate in the incidence of congenital hypothyroidism in Kerman Province, Southeastern Iran. In: The 28th ISEE Conference. September 1–4, 2016, Rome, Italy.

[53] Ranandeh Kalankesh L, Mansouri F, Khanjani N. Association of temperature and humidity with trauma deaths. Trauma Monthly 2015;20(4):e23403. doi:10.5812/traumamon.23403

[54] Linares C, Martinez-Martin P, Rodriguez-Blazquez C, Forjaz MJ, Carmona R, Diaz J. Effect of heat waves on morbidity and mortality due to Parkinson's disease in Madrid: A time-series analysis. Environment International 2016;89–90:1–6. doi:10.1016/j.envint.2016.01.017

[55] Shafa MA, Ebrahimi HA, Khanjani N. A study of the seasonal incidence of multiple sclerosis attacks in Kerman, Iran. Journal of Kerman University of Medical Sciences 2014;21(5):376–83.

[56] Bobb JF, Obermeyer Z, Wang Y, Dominici F. Cause-specific risk of hospital admission related to extreme heat in older adults. JAMA 2014;312(24):2659–67. doi:10.1001/jama.2014.15715

[57] Shajari A, Sanjerehei MM. Modeling the distribution of urolithiasis prevalence under projected climate change in Iran. Urolithiasis 2015;43:339–47. doi:10.1007/s00240-015-0784-2

[58] Marques A, Nunes ML, Moore SK, Strom MS. Climate change and seafood safety: Human health implications. Food Research International 2010;43:1766–79. doi:10.1016/j.foodres.2010.02.010

[59] Davis CE. Globalization, climate change, and human health. The New England Journal of Medicine 2013;369(1):94–5. doi:10.1056/NEJMc1305749#SA1

[60] Zeng W, Lao X, Rutherford S, Xud Y, Xu X, Lina H, et al. The effect of heat waves on mortality and effect modifiers in four communities of Guangdong Province, China. Science of the Total Environment 2014;482–483:214–21.

[61] El-Zein A, Tewtel-Salem M. On the association between high temperature and mortality in warm climates. Science of the Total Environment 2005;343:273–75.

Chapter 5

# Climate Model Sensitivity with Respect to Parameters and External Forcing

Sergei A. Soldatenko and Denis Chichkine

Additional information is available at the end of the chapter

http://dx.doi.org/10.5772/64710

**Abstract**

Mathematical modelling is one of the most powerful methods for the study and understanding of the Earth's climate system and its components. Modern climate models used in variety of applications are derived from a set of multi-dimensional non-linear differential equations in partial derivatives, which describe dynamical, physical and chemical processes and cycles in the climate system. Climate models are mostly deterministic with a large-phase space dimension containing a vast number of parameters that have various meanings. Most of them are not well-known *a priori* and, hence, are not well defined. Parameter errors and their time and space variabilities generate parametric uncertainty. Some model parameters describe external forcing that can strongly influence the climate model behaviour. It is, therefore, very important to estimate the influence of variations in parameters on the model behaviour and results of simulations. Questions like these can be answered by performing sensitivity analysis. This chapter considers various methods of sensitivity analysis that can be used: first, to estimate the influence of model parameter variations on its behaviour; second, to identify parameters of climate models and third, to study the model response to external forcing.

**Keywords:** climate, dynamical systems, sensitivity analysis

## 1. Introduction

One of the biggest issues facing humanity today is the observed ongoing global climate change. Consequently, the prediction of future climate as well as changes in climate due to changes in natural processes and human-caused factors (e.g. greenhouse gas emissions) are issues that have deservedly received significant attention. The essential, powerful and effective methodology for solving this class of problems is computational simulation of the Earth's climate system (ECS) and its components with the use of mathematical climate models that can range from relatively simple to fairly complex. Over the past several decades, the use of climate models as an aid in the understanding past, present and future climates has been substantial. The ECS is a natural, extremely complex, large-scale physical system that includes the atmosphere (the Earth's gaseous envelope), hydrosphere (oceans, rivers and lakes), land surface, cryosphere (ice and snow) and finally the biosphere together with lots of natural and anthropogenic cycles (e.g. water cycle, carbon and nitrogen cycles). It is important that the constituent elements of the ECS are characterized by their own specific physical, dynamical and chemical properties [1–6]. Dynamics of the ECS has turbulent nature and displays wave-like fluctuations within a broad time-space spectrum and, therefore, are characterized by high non-linearity [3, 7, 8]. From mathematical viewpoint, the ECS is an extremely sophisticated, interactive, multi-scale, non-linear dynamical system. In this context, climate simulation represents one of the most complex and important applications of dynamical systems theory, its concepts and methods. The instantaneous state of certain characteristics of the ECS (such as temperature, humidity, atmospheric pressure, wind and precipitation) is referred simply to as the weather. Climate, meanwhile, represents the 'average weather' and is characterized by a statistical ensemble of states through which the ECS travels for decades (usually over ~30 years, according to the World Meteorological Organization's definition).

State-of-the-art mathematical climate models used in variety of applications represent systems of multi-dimensional, non-linear differential equations in partial derivatives that are the mathematical statements of basic physical laws, primarily the conservation laws for momentum, mass and energy. Such models also include a variety of empirical and semi-empirical relationships and equations that are based on observations and experience rather than theories. Mathematical climate models are mostly deterministic with a large-phase space dimension, containing a vast number of various parameters. Equations that describe the evolution of the ECS and its processes are quite complicated. Therefore, in the majority of situations, we, unfortunately, cannot solve them analytically with an arbitrary set of initial conditions, even for very simple cases. We can only find an approximate solution using numerical methods such as, for example, Galerkin projection or finite-difference technique. Consequently, climate models have finite space and time resolutions. Due to the limited resolutions of climate models, many physical processes those are very important for climate dynamics cannot be adequately resolved by the model space-time grid and, therefore, should be parameterized, i.e. described parametrically. As a result, the number of model parameters increases significantly. A large number of them have various meanings and are not well-known *a priori* and, hence, are not well defined. Parameter errors and their variabilities in time and space generate parametric uncertainty in mathematical climate models and, undoubtedly, affect the output results. Assessment of the potential impact of variations in climate model parameters on the model behaviour represents essential element of model building and quality assurance. Sensitivity analysis in dynamical systems is a powerful tool that allows us to estimate the influence of model parameters and their variations on the results of computer simulations. It is important to ensure that some model parameters describe external forcing that can strongly influence the climate model behaviour. All human-caused impacts on the ECS (greenhouse gas emissions due to combustion of fossil fuels and industrial processes, aerosol emissions due to biomass burning, changes in albedo due to deforestation, soil tillage and land degradation) can be considered as small external perturbations that are described in climate models via parameter variations. Hence, both equilibrium and transient climate system sensitivity to external forcing can be examined within the framework of sensitivity analysis in dynamical system.

## 2. Elements of dynamical systems theory

Dynamical systems theory [3, 9–11] serves as a very powerful and reliable framework for modelling, studying and predicting the temporal-spatial behaviour of the ECS and its constituent elements. Generally, a certain abstract dynamical system represents a pair $(X, S_t)$, where $X$ is the phase space of a system, and $S_t : X \rightarrow X$ is a family of evolution functions that is parameterized by a real variable $t \in T$. Commonly this variable performs the role of time. It is assumed that the phase space is a complete metric or Banach space that can be either finite- or infinite-dimensional. The set $\gamma_x = \{x(t) : t \in T\}$ is called trajectory (or orbit), where $x(t)$ is continuous function with values in $X$ such that $S_\tau x(t) = x(t+\tau)$ for all $\tau \in T_+$ and $t \in T$. In climate studies, the semi-dynamical systems are of prime interest. Semi-dynamical systems are dynamical systems whose evolution is considered only for $t \geq 0$. For semi-dynamical systems, a family of mappings $S_t$, $t \geq 0$, forms a semi-group that satisfies the following conditions [12]:

1. $S_0 \equiv I$, where $I$ is the identity operator
2. $S_{t+\tau} = S_t \circ S_\tau = S_\tau \circ S_t, \ \forall t,\tau \geq 0$
3. $S_t x$ is continuous in both $t$ and $x \in X$

Continuous-time dynamical system is commonly generated by the set of autonomous ordinary differential equations (ODEs) assuming that $t \in \mathbb{R}_+$

$$\dot{x} = f(x), x(0) = x_0, x \in X \tag{1}$$

Here $x_0 \in X$ is a system state specified at $t = 0$ and $f$ is continuous vector-valued function. The solution of a system (1) $x = x(x, t)$ is determined for all $t \geq 0$ and can be represented as $x(t) = x(x_0, t) = S_t x_0$. However, in climate simulations, we ordinarily deal with discrete dynamical systems. To convert the set of infinite-dimensional differential equations (1) into finite-dimensional discrete form, a few methods can be applied (e.g. finite-difference approximation or spectral projection technique). Consequently, instead of continuous-time dynamical system (1) we can obtain a discrete in time and space dynamical systems that can be solved numerically for given initial conditions:

$$x_{k+1} = f(x_k), \ k \in \mathbb{Z}_+ \tag{2}$$

Given the system state $x_0$ at the initial time $t = 0$, we define the trajectory of $x_0$ under $f$ to be the sequence of points $\{x_k \in X : k \in \mathbb{Z}_+\}$ such that $x_k = f^k(x_0)$, where $f^k$ denotes the $k$-fold composition of $f$ with itself. Note that $f^0(x) \equiv x$ . Eq. (2) uniquely specifies the trajectory of discrete dynamical system if the map $f : X \to X$ and the initial state $x_0$ are given. The state of dynamical system $x_k$ at time $t_k$ is defined by the system state $x_{k-1}$ at the previous time $t_{k-1}$ as $f(x_{k-1})$.

It is important to highlight that climate dynamical systems possess a number of generic properties [3, 9, 12–14]:

First, these systems are non-linear and dissipative. This means that the divergence of corresponding vector field $\nabla \cdot f(x(t))$ is strictly negative and the system's phase volume contracts.

Secondly, from a certain moment of time $t^*$, the norm of the solution for any initial conditions stays bounded: $\| x(t) \| < V_0$ at $t > t^*$, where $V_0$ is the so-called absorbing set in the system phase space. All trajectories of climate dynamical system will ultimately enter to the ball of radius $V_0$. This property guaranties the existence inside $V_0$ of finite-dimensional invariant compact attracting set, which is called the attractor. Attractor is, therefore, a set towards which a system tends to evolve for a wide variety of initial conditions of the system. If starting states of a system are chosen on the attractor, then the corresponding orbits will remain on the attractor. All other trajectories will be attracted to this set fairly fast.

Thirdly, trajectories of climate dynamical systems are generally unstable, exhibiting Lyapunov instability. This means that the $n_\lambda$-dimensional part of phase volume of a system increases along certain directions correspond to $n_\lambda$ positive Lyapunov exponents (note that $n_\lambda < n$).

Finally, a certain unstable trajectory enclosed in a bounded phase volume (attractor) generates a deterministic dynamical chaos, which means that over time, under certain conditions, the behaviour of simulated system begins to resemble a random process, despite the fact that the system is defined by deterministic laws and described by deterministic equations. This phenomenon of deterministic chaos was first uncovered by E. Lorenz as he observed the sensitive dependence of atmospheric convection model output on initial conditions [15]. Consequently, all trajectories that start close enough will diverge from one another, but will never depart from the attractor. The rate of separation of infinitesimally close trajectories is characterized by the positive Lyapunov exponents. The number of directions along which the orbit is unstable is defined by the number of positive Lyapunov exponents $n_\lambda$.

In general, the dynamics of climate system can be conditionally divided into two phases that are correspondingly the motion towards the attractor and the motion along the attractor. When system orbit travels toward the attractor, the system phase volume contracts to the finite-dimensional volume of the attractor. In many theoretical and practical applications, the ECS evolution is considered on its attractor assuming that the system possesses the properties of ergodicity. Then, statistical characteristics of a system (e.g. first $\bar{x} = \langle x \rangle$ and second $\mathrm{var}(x) = \langle x^2 \rangle - \bar{x}^2$ moments) can be calculated by averaging along a certain trajectory. However, attractors of dissipative dynamical systems have highly complex fractal structure. Such attractors are commonly referred to strange attractors. Since the phase volume of dissipative dynamical systems contracts continuously in the limit of large time, the dimension of attractor is found to be smaller than the dimension of system phase space. Since the phase volume of a system is expanded in $n_\lambda$ directions, the dimension of the attractor cannot be less than the dimension of phase space, and the attractor is nested inside a bounded absorbing set. Consequently, the attractor represents a fractal set of dimension $n_A$ nested in the absorbing ball, where $n_\lambda \le n_A < n$. The fractal dimension of attractors of dissipative dynamical systems can be determined by the Kaplan-Yorke conjecture [16]. The so-called Kaplan-Yorke dimension is defined as follows:

$$D_{KY} = j + \frac{\sum_{i=1}^{j} \lambda_i}{|\lambda_{i+1}|} \tag{3}$$

where $j$ is the maximum integer such that the sum of the $j$ largest exponents is still non-negative, i.e. $\sum_{i=1}^{j} \lambda_i > 0$. The sum of all positive Lyapunov exponents, according to the theorem [17], gives an estimate of Kolmogorov-Sinai entropy, i.e. the value showing the mean divergence of the trajectories on the attractors. The arrangement of the Lyapunov exponents in Eq. (3) is as follows: $\lambda_1 \ge \lambda_2 \ge \cdots \ge \lambda_{n_d}$. The multiplicative inverse (reciprocal) of the largest Lyapunov exponent is referred as characteristic $e$-folding time. Note that for chaotic dynamics the characteristic time is finite, while for regular motion it is infinite. As an example, let us consider the barotropic vorticity equation on the rotating Earth:

$$\frac{\partial \nabla \psi}{\partial t} + J(\psi, \nabla \psi + l + h) = \phi - \alpha \nabla \psi + \nu \nabla^2 \psi,\ \psi(0) = \psi_0$$

where $\psi$ is dimensionless geostrophic stream function, $l$ is the Coriolis parameter, $h$ is an orography, $\varphi$ is an external forcing, $\alpha$ and $v$ are friction coefficients, $J(\,,\,)$ is a Jacobian and $\nabla$ is the Laplacian. Barotropic vorticity equation has a finite-dimensional global attractor. The formula for the estimated fractal dimension of this attractor is [12]:

$$D_F \le \left(\frac{12}{\sqrt{\pi}}\right)^{2/3} Gr^{2/3} \left(\frac{1}{2} + \ln\frac{3\sqrt{2}}{\sqrt{\pi}} + \ln G\right)^{1/3}$$

where $Gr = \|\varphi\|_2/(|\lambda_1| v^2)$ is the generalized Grashof number, $\lambda_1$ is the largest ($\lambda_1 < 0$) eigenvalue of the Laplacian of a sphere.

It is necessary to underline that one of the most essential properties of climate dynamical systems is the stability. Generally, the stability of dynamical systems refers to their response to external forcing and other inputs. Dynamical system is considered to be stable if it remains in an 'invariable' state as long as there is no external forcing, and if the system returns to an 'invariable' state when the external forcing is eliminated. The climate system stability may be determined in terms of bounded inputs. In other words, the climate system is stable if every bounded input generates a bounded output. Climate system modelling is concerned not only with the stability of a system but also the degree of its stability. Climate models have a large number of various parameters that may affect the results of computational simulations. Assessment of the climate system response to variations in the model parameters is extremely important and is among the main problems with regards to system stability.

Other important property of climate dynamical system is its predictability. Generally speaking, predictability is the degree to which an accurate prediction of a system's state can be made either qualitatively or quantitatively. It is important to note that pretty much the same models are applied for both weather prediction and climate simulation. Taking into account the chaotic nature of the atmosphere-ocean system, two types of problems associated with predictability can be defined. Predictability of the first kind is associated with numerical weather prediction, which is a Cauchy (initial value) problem. Numerical weather prediction aims to predict, as precisely as possible, the future state of the atmosphere. This problem requires an accurate knowledge of the current state of the atmosphere-ocean system. However, the evolution of this system is highly sensitive to small errors in the initial conditions [15]. Due to the intrinsic limits with regards to the predictability of atmospheric processes, inaccuracies and missing information in the inputs, as well as inadequacies of forecasting models, detailed and useful weather forecasts are limited to about 2 weeks [18].

By contrast, predictability of the second kind focuses on the prediction of the statistical properties of a system with respect to external forcing. Climate simulation belongs to this class of problems and actually represents a boundary value problem that focuses on much longer time scales than numerical weather prediction (typically several months or even years). Boundary conditions (e.g. the energy that reaches the Earth from the Sun, the energy that goes from the Earth back out to space and so on) constrain climate over a long period of time. If these boundary conditions are imposed correctly then we can simulate the Earth's climate in the long run, without paying attention to what the initial conditions are. To control initial instability caused by initial conditions, the spin-up period is usually eliminated from the climate simulation, prior to analyzing the results. Thus, predictability of climate systems involves the study of stability of climate model attractors with respect to external forcing. Predictability of dissipative dynamical systems applied to climate simulations deteriorates with time. To quantify predictability, we can use the rate of divergence of system orbits in phase space (i.e. Kolmogorov-Sinai entropy, Lyapunov exponents).

## 3. A formal dynamic model of climate system

Let us consider the ECS in a bounded spatial-temporal domain $\Omega_t = \Omega \times [0, \tau]$. Let us denote by $\phi \in Q(\Omega_t)$ the state vector that characterizes the ECS in the domain $\Omega_t$. Note that $Q(\Omega_t)$ is the infinite real space of sufficiently smooth state functions that satisfy certain boundary conditions at the boundary $\partial\Omega$ of the spatial domain $\Omega$, which is usually the Earth's sphere.

Let $r \in \Omega \subset \mathbb{R}^3$ be the vector of spatial variables and $t \in [0, \tau]$ be the time. Mathematically, the temporal evolution of the ECS in the domain $\Omega_t$ is expressed via the set of partial differential equations, which reflect the specific dynamical, physical and other properties of the ECS:

$$\partial_t \varphi(r,t) = \mathcal{L}(\varphi(r,t), \lambda(r,t)),\ \varphi(r,0) = \varphi_0(r) \tag{4}$$

where $\mathcal{L}$ is a non-linear, multi-dimensional differential operator that describes the dynamics, dissipation and external forcing of the system, $\lambda \in G(\Omega_t)$ is the parameter vector, $G(\Omega_t)$ is the domain of admissible values of the parameters and $\phi_0$ is a given vector valued function (the initial state estimate). The model state vector $\phi$ includes temperature, pressure, density, humidity, wind velocity and other physical variables. Note that the system (4) characterizes a *continuous medium for which the state vector $\phi$ is infinite-dimensional*: $\phi \in \Phi$, where $\Phi$ is the infinite-dimensional Hilbert space. The vector of parameters, in its turn, contains any input of the system (4) such as classical parameters, initial and boundary conditions and so on. Essentially, the solution of such complex infinite-dimensional system cannot be found analytically. In order to obtain numerical solution, the original system of infinite-dimensional equations should be transformed, using an appropriate method, into a system with a finite number of degrees of freedom. For example, Eq. (4) can be projected onto the sub-space spanned by the orthogonal base $\Psi = \{\psi_i\}_{i=1}^{n}$ that is defined to represent state vector on the domain $\Omega$. Thus, state vector $\phi$ can be introduced in the form of normally convergent series:

$$\varphi(r,t) \approx \sum_{i=1}^{n} x_i(t)\psi_i(r) \tag{5}$$

Substituting (5) into (4) and then applying the Galerkin method, we can obtain, instead of infinite-dimensional distributed parameter system (4), the finite-dimensional lumped system that is formally described by the following set of ordinary differential equations:

$$\dot{x} = F(x,\alpha),\ t \in [0,\tau],\ x(0) = x_0 \tag{6}$$

Here $x \in \mathbb{R}^n$ is the state vector with dimension $n$ representing a set of spectral coefficients, $F$ is the Galerkin projection of the operator $\mathcal{L}$ on the base $\Psi$ and $\alpha \in \mathbb{R}^m$ is the $m$-dimensional parameter vector.

It is important to note that space-time spectrum of processes occurring in the climate system is extremely broad. Consequently, the state-of-the-art mathematical climate models due to their spatial-temporal limited resolutions are unable to simulate correctly all of these processes. Physical processes that are too small-scale to be explicitly represented in the model due to its discrete spatial-temporal structure are parameterized, i.e. replaced by simplified parametric schemes generating additional model parameters. Let $\tau_c$ be a characteristic timescale of a certain physical process. Physical processes with timescales smaller than $\tau_c$ (the so-called sub-grid-scale processes) should be parameterized in climate models. Examples of these processes include radiative transfer (short-wave solar radiation and outgoing long-wave radiation emitted by the Earth), cloud formation processes, microphysical processes within clouds that lead to the formation of precipitation, the deep convection in the tropics, land-surface processes, photochemical processes, carbon cycle, etc.

Using finite difference method to approximate time derivatives in Eq. (6), we can obtain the generalized numerical climate model that can be used for computer simulations:

$$x_k = m_{0,k}\left(x_0\right), \quad k = 1,\ldots,K$$

where $m_{0,k}$ is non-linear operator that indirectly contains model parameters and propagates the state vector from time $t_0$ (the initial conditions) to time $t_k$, and $K$ is the number of time steps. Generally, all climate models may be arranged in several classes, based on various principles, e.g. the complexity of the models or the description and representation of physical processes [19]. However, there is no best or general-purpose climate model. Each particular model is characterized by inherent properties, and has specific advantages and disadvantages. Selecting the 'best' model depends on many various factors, including the objectives of simulation and what performance measures are used.

## 4. Forward and adjoint sensitivity analysis

To estimate the impact of model parameter variations on the model performance and state variables, one can use a sensitivity function (or coefficient) that is the partial derivative of a given element of state vector $x_i$ with respect to a certain model parameter $\alpha_j$ [20, 21]:

$$S_{ij}\left(t,\alpha\right) \equiv \left.\frac{\partial x_i\left(t,\alpha\right)}{\partial \alpha_j}\right|_{\alpha^0} = \lim_{\delta\alpha_j \to 0} \frac{x_i\left(t,\alpha_j^0 + \delta\alpha_j\right) - x_i\left(t,\alpha_j^0\right)}{\delta\alpha_j} \tag{7}$$

where $\delta\alpha_j$ is the infinitesimal perturbation of parameter $\alpha_j$ around some fixed point $\alpha_j^0$. Approximating the state vector $x(\alpha^0 + \delta\alpha)$ around $x(\alpha^0)$ by Taylor expansion, one can obtain the following linear equation:

$$x\left(\alpha^0 + \delta\alpha\right) = x\left(\alpha^0\right) + \left.\frac{\partial x}{\partial \alpha}\right|_{\alpha^0} \delta\alpha + \text{H.O.T.}$$

where $\left.\frac{\partial x}{\partial \alpha}\right|_{\alpha^0} \equiv S_\alpha \in \mathbb{R}^{n \times m}$ is a sensitivity matrix. Let us rewrite the vector equation (6) in component form:

$$\dot{x}_i = F_i(x,\alpha), t \in [0,\tau], x_i(0) = x_{i0}, i \in [1,n] \tag{8}$$

Differentiating Eq. (8) with respect to $\alpha_j$, we obtain the set of non-homogeneous ODEs, the so-called sensitivity equations:

$$\frac{d}{dt}\left(\frac{\partial x_i}{\partial \alpha_j}\right) = \sum_{k=1}^{n}\left(\frac{\partial F_i}{\partial x_k}\frac{\partial x_k}{\partial \alpha_j} + \frac{\partial F_i}{\partial \alpha_j}\right), i \in [1,n], j \in [1,m] \tag{9}$$

Sensitivity equations describe the evolution of sensitivity functions along a given trajectory, and, therefore, allow tracing the sensitivity dynamics in time. A system of sensitivity equations (9) can be expressed in a matrix form (see below). Thus, to calculate sensitivity functions with respect to parameter $\alpha_j$ one should be able to solve the following set of differential equations with given initial conditions:

$$\begin{cases} \dot{x}_i = F_i(x,\alpha), & x_i(0) = x_{i0} \\ \dot{S}_j = M \cdot S_j + D_j, & S_j(0) = S_{j0} \end{cases} \tag{10}$$

where $S_j = (\partial x/\partial \alpha_j) = (S_{1j}, \cdots, S_{nj})^{\mathrm{T}}$ is the sensitivity vector with respect to parameter $\alpha_j$, $M \equiv M_i^j = \left(\partial F_i / \partial x_j\right)$ is a Jacobian matrix and $D_j = (\partial F_1/\partial \alpha_j, \partial F_2/\partial \alpha_j, \ldots, \partial F_n/\partial \alpha_j)^{\mathrm{T}}$. Once we have solved Eq. (10), it is possible to analyse the sensitivity of system (8) with respect to the parameter $\alpha_j$. Since the model parameter vector $\alpha$ has $m$ components, to evaluate the model response to variations in the parameter vector $\delta\alpha$, the set of Eq. (10) must be solved $m$ times. Therefore, this approach is acceptable for low-order models. The use of sensitivity functions requites the differentiation of model equations with respect to parameters. However, this is not always possible. Fairly often in sensitivity analysis of complex dynamical systems, the model response to variations in its parameters represents a generic response function that characterizes the dynamical system [22, 23]:

$$R(x,\alpha) = \int_0^\tau \Phi(t;x,\alpha)\,dt \tag{11}$$

where $\Phi$ is a non-linear function of state variables $x$ and model parameters $\alpha$. The gradient of functional $R$ with respect to vector of parameters $\alpha$ around the unperturbed parameter vector $\alpha^0$ and corresponding unperturbed state vector $x^0$:

$$\nabla_\alpha R\left(x^0,\alpha^0\right)=\left(\frac{dR}{d\alpha_1},\ldots,\frac{dR}{d\alpha_m}\right)^{\mathrm{T}}\Bigg|_{x^0,\alpha^0}$$

quantifies the influence of parameters on the model performance. In particular, the effect of the $j$th parameter can be estimated as follows:

$$\left.\frac{dR}{d\alpha_j}\right|_{\alpha_j^0}\approx\frac{R\left(x^0+\delta x;\alpha_1^0,\ldots,\alpha_j^0+\delta\alpha_j,\ldots,\alpha_m^0\right)-R\left(x^0,\alpha^0\right)}{\delta\alpha_j}$$

where $\delta\alpha_j$ is the variation in parameter $\alpha_j^0$. Note that

$$\frac{dR}{d\alpha_j}=\sum_{i=1}^{n}\frac{\partial R}{\partial x_i}\frac{\partial x_i}{\partial\alpha_j}+\frac{\partial R}{\partial\alpha_j}=\sum_{i=1}^{n}S_{ij}\frac{\partial x_i}{\partial\alpha_j}+\frac{\partial R}{\partial\alpha_j}$$

The accuracy of sensitivity estimates strongly depends on the choice of perturbation $\delta\alpha_j$. By introducing the Gâteaux differential, the sensitivity analysis problem can be considered in the differential formulation eliminating the need to set the value of $\delta\alpha_j$ [22, 23]. The Gâteaux differential for the response function (11) has the following form:

$$\delta R\left(x^0,\alpha^0\right)=\int_0^\tau\left(\left.\frac{\partial\Phi}{\partial x}\right|_{x^0,\alpha^0}\cdot\delta x+\left.\frac{\partial\Phi}{\partial\alpha}\right|_{x^0,\alpha^0}\cdot\delta\alpha\right)dt \tag{12}$$

Here, $\delta x$ is the state vector variation due to the variation in the parameter vector in the direction $\delta\alpha$. Linearizing the non-linear model (Eq. (8)) around an unperturbed trajectory $x^0(t)$, we obtain the following system of variational equations, the so-called tangent linear model, for calculating $\delta x$:

$$\frac{\partial\delta x}{\partial t}=\left.\frac{\partial F}{\partial x}\right|_{x^0,\alpha^0}\cdot\delta x+\left.\frac{\partial F}{\partial\alpha}\right|_{x^0,\alpha^0}\cdot\delta\alpha,t\in\left[0,\tau\right],\delta x\left(0\right)=\delta x_0$$

Then, using Eq. (12), we can calculate the variation in the response functional $\delta R$. Taking into account that $R(x^0,\alpha^0;\delta x,\delta\alpha)=\langle\nabla_\alpha R,\delta\alpha\rangle$, where $\langle\cdot,\cdot\rangle$ is a dot-product, the model sensitivity to variations in the parameters is estimated via the gradient of the response functional $\nabla_\alpha R$. This approach is convenient, however, computationally very expensive since climate models involve a large number of parameters. Adjoint approach allows the calculation of sensitivities within a single numerical experiment. Using adjoint equations, one can calculate the sensitivity gradient $\nabla_\alpha R$ as follows [22]:

$$\nabla_\alpha R\left(x^0,\alpha^0\right) = \int_0^\tau \left[ \left.\frac{\partial \Phi}{\partial \alpha}\right|_{x^0,\alpha^0} - \left( \left.\frac{\partial F}{\partial \alpha}\right|_{x^0,\alpha^0} \right)^{\mathrm{T}} \cdot x^* \right] dt \tag{13}$$

where the vector function $x^*$ is the solution of adjoint model, which is numerically integrated in the inverse time direction:

$$-\frac{\partial x^*}{\partial t} - \left( \left.\frac{\partial F}{\partial x}\right|_{x^0,\alpha^0} \right)^{\mathrm{T}} \cdot x^* = -\left.\frac{\partial \Phi}{\partial x}\right|_{x^0,\alpha^0}, t \in [0,\tau], x^*(0) = 0 \tag{14}$$

One of the key problems in climate simulation is the identification of parameters in mathematical models that describe the climate system. This problem is quite difficult due to both the huge number of state variables and parameters, and the argument that the governing finite-difference equations are non-linear grid functions of these states and parameters. Using the adjoint approach, we can solve the identification parameter problem if the observations are available. This problem is mathematically formulated as an optimal control problem in which model parameters play the role of control variables, and model equations are considered as constraints. Let $y^{obs}$ be the set of observations and $H$ be observation operator mapping from solution space of model to observation space. Therefore, $y^{obs} = H(x) + \varepsilon^{obs}$, where $\varepsilon^{obs}$ is the vector of observation errors. It is usually assumed that these errors are serially uncorrelated and normally distributed with known covariance matrix $W$. The parameter identification problem seeks to minimize, with respect to $\alpha$, a certain objective function $\mathcal{J}(x, \alpha)$ expressing the 'distance' between observations and corresponding model state using the model equations as constraints:

$$\alpha^* = \arg\min \mathcal{J}(x,\alpha)$$

where $\alpha^*$ is a specified parameter vector. The objective function is written as

$$J(\alpha) = \frac{1}{2}\left\| H(x) - y^{obs} \right\|_R^2 \tag{15}$$

For illustrative purposes, let us consider the following example. Let $y_i^{\text{obs}}(t_k)$ and $x_i(t_k, \alpha)$ be, respectively, the observation and model prediction of the $i$th component of state vector at time $t_k$, $\sigma_i^2$ the variance of $y_i^{\text{obs}}(t_k)$, and $H$ the identity operator. Then the objective function can be written as

$$\mathcal{J}(x,\alpha)=\frac{1}{2}\sum_{(i)}\sum_{(k)} w_i\left[x_i(t_k,\alpha)-y_i^{obs}(t_k)\right]^2 \tag{16}$$

where $w_i$ are a *weighted coefficient reflecting the accuracy of* observations (in our case, $w_i = 1/\sigma_i^2$). Many optimization algorithms rely on descent methods that require the computation of the gradient of the objective function. The gradient of Eq. (16) with respect to parameter $\alpha_j$ is defined as

$$\frac{\partial \mathcal{J}}{\partial \alpha_j}=-\sum_{(i)}\sum_{(k)} w_i \cdot x_i(t_k)\frac{\partial x_i(t_k,\alpha)}{\partial \alpha_j} \tag{17}$$

where $\Delta x_i(t_k) = x_i(t_k, \alpha) - y_i^{\text{obs}}(t_k)$. The right-hand side of Eq. (17) shows that sensitivity functions play a critical role in determining the corrected values of model parameters.

## 5. Application of conventional methods of sensitivity analysis

Both forward and adjoint methods allow us to analyse transient and equilibrium sensitivities of dynamical systems. The exploration of sensitivity of complex dynamical systems requires considerable computational resources. For simple enough low-order models, the computational cost is minor and, for that reason, models of this class are widely used as simple test instruments to emulate more complex systems. We will illustrate the above-described conventional methods of sensitivity analysis based on a coupled non-linear dynamical system, which is composed of fast (the 'atmosphere') and slow (the 'ocean') versions of the well-known Lorenz [15] model (L63). This model allows us to mimic the atmosphere-ocean system and therefore serves as a key element of a theoretical and computational framework for the study of various aspects of sensitivity analysis. Recall that under certain conditions the Lorenz model exhibits a chaotic behaviour. As mentioned above, the system is obtained by coupling of two versions of the original Lorenz model with distinct timescales that differ by factor $\varepsilon$ [24, 25]:

$$\left.\begin{array}{c}\dot{x} = \sigma(y - x) - c(aX + k) \\ \dot{y} = rx - y - xz + c(aY + k) \\ \dot{z} = xy - bz + c_z Z\end{array}\right\} \quad (18)$$

$$\left.\begin{array}{c}\dot{X} = \varepsilon\sigma(Y - X) - c(x + k) \\ \dot{Y} = \varepsilon(rX - Y - aXZ) + c(y + k) \\ \dot{Z} = \varepsilon(aXY - bZ) - c_2 z\end{array}\right\} \quad (19)$$

where the fast and slow sub-systems are represented, respectively, by lower case and capital letters. The following notations are used in Eqs. (16)–(21): $\sigma$, $r$ and $b$ are the parameters of the original L63 model, $a$ is a parameter representing the amplitude scale factor, $k$ is 'uncentring' parameter, $c$ is a coupling strength parameter for $x - X$ and $y - Y$ variables, and $c_z$ is a coupling strength parameter for $z - Z$ variables. One can assume that $a = 1$, $k = 0$ and $c = c_z$ without loss of generality. Therefore, the vector of state variables of coupled model (18) and (19) is $x = (x, y, z, X, Y, Z)^{\mathrm{T}}$ and the vector of model parameters is $\alpha = (\sigma, r, b, c, \varepsilon)^{\mathrm{T}}$. In the operator form, the set of Eqs. (16)–(21) can be rewritten as follows:

$$\dot{x} = (L + Q)x \quad (20)$$

where the non-linear uncoupled operator $L$ and linear coupled operator $Q$ are represented by the following matrices:

$$L = \begin{bmatrix} -\sigma & \sigma & 0 & & & \\ r & -1 & -x & & 0 & \\ 0 & x & -b & & & \\ & & & -\varepsilon\sigma & \varepsilon\sigma & 0 \\ & 0 & & \varepsilon r & -\varepsilon & -\varepsilon X \\ & & & 0 & \varepsilon X & -\varepsilon b \end{bmatrix}, Q = \begin{bmatrix} & & & -c & 0 & 0 \\ & 0 & & 0 & c & 0 \\ & & & 0 & 0 & c \\ -c & 0 & 0 & & & \\ 0 & c & 0 & & 0 & \\ 0 & 0 & -c & & & \end{bmatrix}$$

The unperturbed parameter values are selected to be as follows:

$$\sigma^0 = 10, r^0 = 28, b^0 = 8/3, \varepsilon^0 = 0.1, c^0 \in [0.1; 1.2]$$

Chosen values of $\sigma$, $r$ and $b$ correspond to chaotic behaviour of the L63 model. For $\sigma = 10$ and $b = 8/3$, the critical value of parameter $r$ is 24.74, which means that any value of $r$ larger than 24.74 induces deterministic chaos [15]. The parameter $\varepsilon = 0.1$ indicates that the slow system is 10 times slower than the fast system. Basic dynamical, correlation and spectral properties of this system, and the influence of the coupling strength on power spectrum densities, spectrograms and autocorrelation functions were explored in detail in [26]. Here, we will briefly mention the main features of the system (19).

The coupling strength parameter $c$ plays a very important role in qualitative changes in the system dynamics since this parameter controls the interactions between fast and slow subsystems. Qualitative changes in the dynamical properties of a system can be revealed by determining and analyzing the system's spectrum of Lyapunov exponents that characterize the average rate of exponential divergence (or convergence) of nearby trajectories in the phase space. In the analysis of coupled dynamical systems, we are dealing with conditional Lyapunov exponents that are normally used to characterize the synchronization with coupled systems. The dynamical system (21) has six distinct Lyapunov exponents. If the parameter $c$ tends to zero, then this system has two positive, two zero and two negative Lyapunov exponents. Let us examine the impact of coupling strength parameter on the two largest Lyapunov exponents (**Figure 1**). These exponents, being initially positive, monotonically decrease when the coupling strength parameter increases approaching the $x$-axis at about $c \approx 0.8$ and become negative at $c \approx 0.95$. Therefore, phase-synchronous regime for fast and slow sub-systems is observed when $c > 0.95$ [27]. However, when $c > 1.0$ , all six Lyapunov exponents become negative causing a limit cycle dynamics.

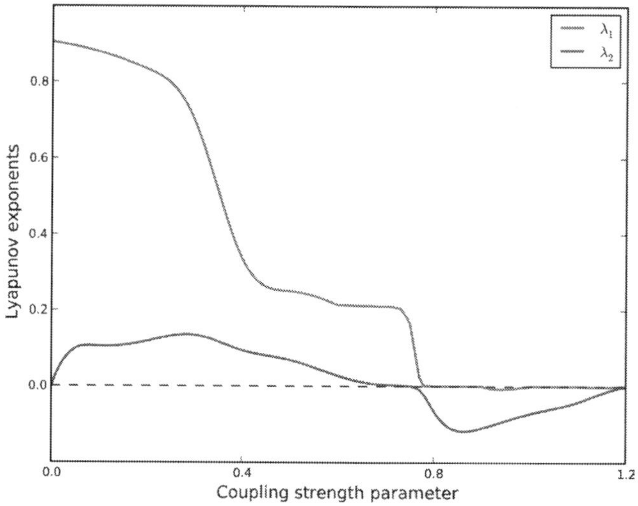


**Figure 1.** Two largest conditional Lyapunov exponents as functions of coupling strength parameter.

Apart from Lyapunov exponents, autocorrelation functions enable one to distinguish between regular and chaotic processes and to detect transition from order to chaos. In particular, for chaotic motions, autocorrelation functions decrease in time, in many cases exponentially, while for regular motions, autocorrelation functions are unchanged or oscillating. In general, however, the behaviour of autocorrelation functions for chaotic motions is frequently very complicated and depends on many factors (e.g. [28]). Knowing autocorrelation functions, one can determine a typical timescale (typical time memory) of the process [29]. Moreover, if autocorrelation functions are positive, the dynamical system may have the persistence property (an intention of a system to remain in the similar state from one time moment to the following). For a given discrete dynamic variable $\{x_i\}_{i=0}^{n}$, an autocorrelation function is determined as $c(s) = \langle x_i x_s \rangle - \langle x_i \rangle \langle x_s \rangle$, where the brackets $\langle \cdot \rangle$ denote ensemble averaging. Assuming time series originates from a stationary and ergodic process, ensemble averaging can be replaced by time averaging over a single normal realization $c(s) = \langle x_i x_s \rangle - \langle x \rangle^2$.

Signal analysis commonly uses the normalized autocorrelation function (ACF), defined as $R(s) = c(s)/c(0)$. Results of numerical experiments show that for relatively small parameter $c$ ($c < 0.4$), the ACFs and their envelopes for all variables decrease fairly rapidly to zero, consistently with the chaotic behaviour of the coupled system. However, as expected, the rates of decay of the ACFs of the slow variables are less than that of the fast variables. For coupling strength parameter on the interval $0.4 < c < 0.6$, the ACFs of the fast variables become smooth and converge to zero. As the parameter $c$ increases, the ACFs become periodic and their envelopes decay slowly with time, indicating transition to regularity. For $c > 0.8$, calculated ACFs show periodic signal components. In order to explore the sensitivity of system (19) with respect to coupling strength parameter, let us introduce the following sensitivity functions:

$$S_{1c} = \partial x / \partial c, S_{2c} = \partial y / \partial c, S_{3c} = \partial z / \partial c$$

$$S_{4c} = \partial X / \partial c, S_{5c} = \partial Y / \partial c, S_{6c} = \partial Z / \partial c$$

The corresponding sensitivity equations can be written as

$$\dot{S}_{1c} = \sigma\left(S_{2c} - S_{1c}\right) - cS_{4c} - X$$

$$\dot{S}_{2c} = rS_{1c} - S_{2c} - xS_{3c} - zS_{1c} + cS_{5c} + Y$$

$$\dot{S}_{3c} = xS_{2c} + yS_{1c} - bS_{3c} + cS_{6c} + Z$$

$$\dot{S}_{4c} = \varepsilon\sigma\left(S_{5c} - S_{4c}\right) - cS_{1c} - x$$

$$\dot{S}_{5c} = \varepsilon\left(rS_{4c} - S_{5c} - XS_{6c} - ZS_{4c}\right) + cS_{2c} + y$$

$$\dot{S}_{6c} = \varepsilon\left(XS_{5c} + YS_{4c} - bS_{6c}\right) - cS_{3c} - z$$

Sensitivity functions can be introduced for any particular model parameter. Since the parameter vector $\alpha$ consists of five components, five sets of sensitivity equations can be derived from the model (Eq. (19)). The dynamics of sensitivity functions can be traced by solving the sensitivity equations along with the non-linear model. Sensitivity functions, calculated on the time interval [0, 20], are shown in **Figure 2**. Envelopes of these functions grow over time while sensitivity functions themselves oscillate. Sensitivity function is a measure of the change in state variable due to the variation in the estimated parameter. Unfortunately, obtained sensitivity functions are inherently uninformative and misleading. We cannot make a clear conclusion from them about system sensitivity to variations in the parameter $c$. Similar results were obtained when we considered the influence of variations in the parameter $r$ on the system dynamics. This parameter plays an important role in the formation of system's dynamical structure and transition to chaos. Let us define the following sensitivity functions:

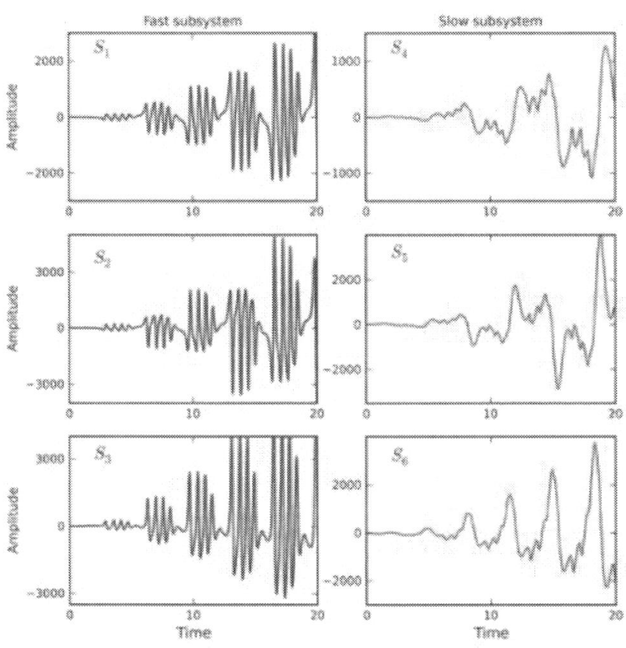


**Figure 2.** Time dynamics of sensitivity functions with respect to parameter $c$ on the time interval [0, 20] for $c^0 = 0.9$.

$$S_{1r} = \partial x / \partial r, S_{2r} = \partial y / \partial r, S_{3r} = \partial z / \partial r$$

$$S_{4r} = \partial X / \partial r, S_{5r} = \partial Y / \partial r, S_{6r} = \partial Z / \partial r$$

The associated system of sensitivity equations can be written as

$$\dot{S}_{1r} = \sigma(S_{2r} - S_{1r}) - cS_{4r}$$

$$\dot{S}_{2r} = x + rS_{1r} - S_{2r} - (xS_{3r} + zS_{1r}) + cS_{5r}$$

$$\dot{S}_{3r} = (xS_{2r} + yS_{1r}) - bS_{3r} + cS_{6r}$$

$$\dot{S}_{4r} = \varepsilon\sigma(S_{5r} - S_{5r}) - cS_{1r}$$

$$\dot{S}_{5r} = \varepsilon\left[X + rS_{4r} - S_{5r} - (XS_{6r} + ZS_{4r})\right] + cS_{2r}$$

$$\dot{S}_{6r} = \varepsilon\left[(XS_{5r} + YS_{4r}) - bS_{6r}\right] - cS_{3r}$$

Envelopes of calculated sensitivity functions grow over time and sensitivity functions demonstrate the oscillating behaviour (**Figure 3**). Obtained functions are also uninformative and inconclusive. Thus, using conventional methods of sensitivity analysis can be questionable in terms of interpretation of the obtained results for chaotic dynamics. As discussed in [21], general solutions of sensitivity equations for oscillatory non-linear dynamical systems grow unbounded as time tends to infinity; therefore, sensitivity functions calculated by conventional approaches have a high degree of uncertainty, quickly becoming uninformative and inconclusive as time increases. In this regard, in climate simulation, the average values of sensitivity functions $\nabla_\alpha\langle R(\alpha)\rangle$ over a certain period of time can be considered as one of the most important measures of sensitivity, where $R$ is a generic response functional (Eq. (11)). However, the gradient $\nabla_\alpha\langle R(\alpha)\rangle$ cannot be correctly estimated within the framework of conventional methods of sensitivity analysis since for chaotic systems it is observed [30–32] that $\nabla_\alpha\langle R(\alpha)\rangle \neq \langle\nabla_\alpha\langle R(\alpha)\rangle$ . This is because the integral

$$I = \lim_{\tau\to\infty}\int_0^\tau \lim_{\delta\alpha\to 0}\frac{R(\alpha+\delta\alpha) - R(\alpha)}{\delta\alpha}dt$$

does not possess uniform convergence and two limits ($\tau \to \infty$ and $\delta\alpha \to 0$) would not commute.

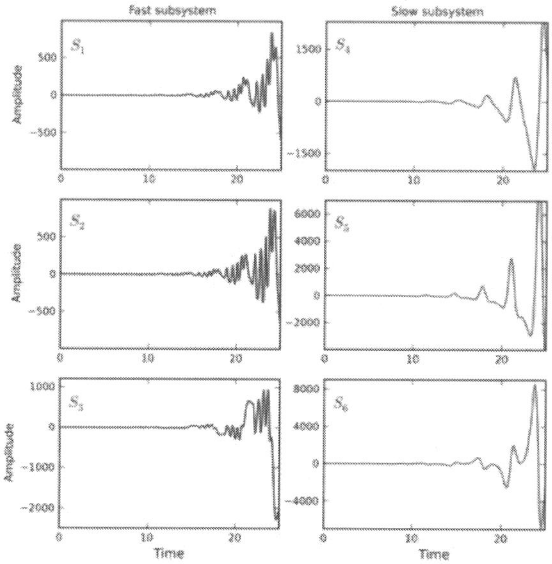


**Figure 3.** Time dynamics of sensitivity functions with respect to parameter $r$ on the time interval [0, 25] for $c^0 = 0.9$.

## 6. Sensitivity analysis based on shadowing property

The new approach of sensitivity analysis known as shadowing method was introduced in [31] to analyse the sensitivity of highly non-linear and/or chaotic dynamical systems with respect to variations in their parameters. This method is theoretically based on the pseudo-orbit tracing (or shadowing) property for discrete and continuous dynamical systems [33, 34]. Using shadowing method, one can properly estimate the time-average sensitivities $\langle\nabla_\alpha\langle R(\alpha)\rangle\rangle$ making the conclusion on the system sensitivity to variations in its parameters. The pseudo-orbit tracing property means that around an approximate (or pseudo) trajectory of dynamical system under consideration, there exists the exact trajectory, such that it locates evenly close to a pseudo-trajectory. Pseudo-trajectories arise by virtue of various errors of computer simulation (e.g. round-off errors and numerical technique errors). Thus, by making numerical simulations, we actually cannot obtain an exact trajectory of a system, but only an approximate trajectory known as a pseudo-trajectory. The exploration of shadowing property in dynamical systems was originated by Anosov [35] and Boven [36]. To date, the shadowing theory is well-established for the so-called hyperbolic dynamics that distinguished by the existence of both expanding and contracting lines for derivatives. Let us make some basic notes on the shadowing theory.

Let $f: X \to X$ be a homeomorphism of a compact metric space ($X$, dist). We will consider the discrete dynamical system $\mathcal{F}: \mathbb{Z} \times X \to X$ generated by homeomorphism $f$:

$$\mathcal{F}(k,x) = f^k(x),\ k \in \mathbb{Z} \text{ and } x \in X.$$

Let us denote by $o(x, f)$, the trajectory of a point $x \in X$ with respect to $f$, i.e. $O(x,f) = \{f^k(x): k \in \mathbb{Z}\}$. Set any metric (distance function) $dist(\cdot,\cdot)$ for X. It is said that a set of points $\{x_k: k \in \mathbb{Z}\}$ is called a $d$-pseudo-trajectory ($d > 0$) of $f$ if $dist(x_{k+1}, f(x_k)) < d$ for $k \in \mathbb{Z}$. Note that a distance function $dist(\cdot,\cdot)$ defines an interval between each pair of geometric objects inside the brackets. It is said that the dynamical system $f$ possesses the shadowing property if for every $\varepsilon > 0$, there is $d > 0$ such that for every $d$-pseudo-trajectory $\zeta$, there exists $y \in X$ satisfying $dist(f^k(y), x_k) < \varepsilon$ for all $k \in \mathbb{Z}$. According to the so-called discrete shadowing lemma [33], for any $\varepsilon > 0$, there exists $d > 0$, such that any $d$-pseudo-trajectory can be $\varepsilon$-shadowed. If any dynamical system has the shadowing property, its trajectories calculated numerically reflect the reality.

The shadowing lemma for continuous dynamical systems (flows) is more sophisticated than for discrete systems since in this instance re-parameterization of shadowing trajectories is required because for flows, close points of the true trajectory and the pseudo orbit do not correspond to the same time moments [33, 34]. Let $X$ be the phase space of continuous dynamical system $f^t: \mathbb{R} \times X \to X$ generated by a set of ordinary differential equations. A function $g: \mathbb{R} \to X$ is a $d$-pseudo-orbit of $f^t$ if the inequality $dist(f^t(t, g(\tau)), g(t + \tau)) < d$ holds for any $\tau \in \mathbb{R}$ and $t \in [-1, 1]$. A re-parameterization is actually a monotonically increasing homeomorphism $h$ of the line $\mathbb{R}$ such that $h(0) = 0$. The set of re-parameterizations $h$ denoted by $Rep(\varepsilon)$, where $\varepsilon > 0$, is defined as [33]:

$$Rep(\varepsilon) = \left\{ h \in Rep \left| \frac{h(t_1) - h(t_2)}{t_1 - t_2} \right| \leq \varepsilon \right\} \qquad \text{for any different } t_1, t_2 \in \mathbb{R}$$

To illustrate the applicability of this method, let us consider a continuous one-parameter generic dynamical system

$$\dot{x} = f(x, \alpha),\ x \in \mathbb{R}^n,\ \alpha \in \mathbb{R} \tag{21}$$

The shadowing sensitivity analysis method is based on the 'continuous' shadowing lemma and the following assumptions: (a) the dynamical system is ergodic and (b) model state variables are considered over long time interval $t \in [0, T]$, where $T \to \infty$, and an averaged performance measure $\langle R(\alpha) \rangle = \lim_{T \to \infty} \frac{1}{T} \int_0^T R(x(t,\alpha),\alpha)dt$ is of the most interest for us. With these assumptions, we can use the arbitrarily chosen trajectory of the system to trace the state variables along the orbit and calculate $R(\alpha)$. For example, the arbitrary trajectory $x(t)$ can be chosen as a solution of the model equations, such that it is located nearby a certain reference trajectory $x_r(t)$. According to the shadowing lemma, the closest orbit $x(t)$ to $x_r(t)$ satisfies the following constrained minimization problem [37]:

$$\min_{x,\tau} \frac{1}{T} \int_0^T \left[ \left\| x(\tau(t)) - x_r(t) \right\|^2 + \eta^2 \left( \frac{d\tau}{dt} - 1 \right)^2 \right] dt \text{ such that } \frac{dx}{dt} = f(x,\alpha), \tag{22}$$

where $\eta$ is the parameter that provides the same order of magnitude of the two terms in the integrand and $\tau(t)$ is a time transformation. The second term in the integrand describes re-parameterization. The problem (22) is called the non-linear least square shadowing (LSS) problem, and its solution denoted by $x_S^{(T)}(t,\alpha)$ and $\tau_S^{(T)}(t,\alpha)$ is a solution of the model equations and time transformation that provides the trajectory $x_S^{(T)}(t,\alpha)$ to be close to $x_r(t)$. The performance measure $\langle R(\alpha) \rangle$ averaged over the time interval $t \in [0, T]$ can be then approximated as:

$$R(\alpha) \approx R_S^{(T)}(\alpha) = \frac{1}{\tau(T) - \tau(0)} \int_{\tau(0)}^{\tau(T)} R\left( x_S^{(T)}(t,\alpha), \alpha \right) dt \tag{23}$$

since $x_S^{(T)}(t,\alpha)$ satisfies the model equation at a different $\alpha$. The desired sensitivity estimate $\nabla_\alpha \left\langle R_S^{(T)}(\alpha) \right\rangle$ can be computed by solving the following linearized LLS problem [37]:

$$\min_{S,\mu} \frac{1}{T} \int_0^T \left[ \|S\|^2 + \eta^2 \mu^2 \right] dt \text{ such that } \frac{dS}{dt} = \frac{\partial f}{\partial x} S + \frac{\partial f}{\partial \alpha} + \mu f(x_r, \alpha), \tag{24}$$

The solutions of this problem $S(t)$ and $\mu(t)$ relate to the solutions of the non-linear LSS problem (22) as follows:

$$S(t) = \frac{d}{d\alpha} \left[ x_S^{(T)} \left( \tau_S^{(T)}(t,\alpha), \alpha \right) \right], \quad \mu(t) = \frac{d}{d\alpha} \frac{d\tau_S^{(T)}(t,\alpha)}{dt} \tag{25}$$

The time-dependent parameter $\mu$ is called a time-dilation variable, and it corresponds to the time transformation from the shadowing lemma. Using $S$ and $\mu$, we can estimate the desired sensitivity (the derivative of the linearized performance measure (22) with respect to the parameter $\alpha$):

http://dx.doi.org/10.5772/64710

$$\nabla_{\alpha}\left\langle R_{s}^{(T)}(\alpha)\right\rangle \approx \frac{1}{T}\int_{0}^{T}\left(\frac{\partial R}{\partial x}S+\frac{\partial R}{\partial \alpha}+\mu\left(R-\bar{R}\right)\right)dt,\ \text{where}\ \bar{R}=\frac{1}{T}\int_{0}^{T}Rdt \tag{26}$$

Several numerical algorithms can be used to solve the linearized LSS problem (24). One such method is based on variational calculus, which is used to derive optimality conditions representing a system of linear differential equations that are then discretized and solved numerically to calculate variables $S$ and $\mu$ [37].

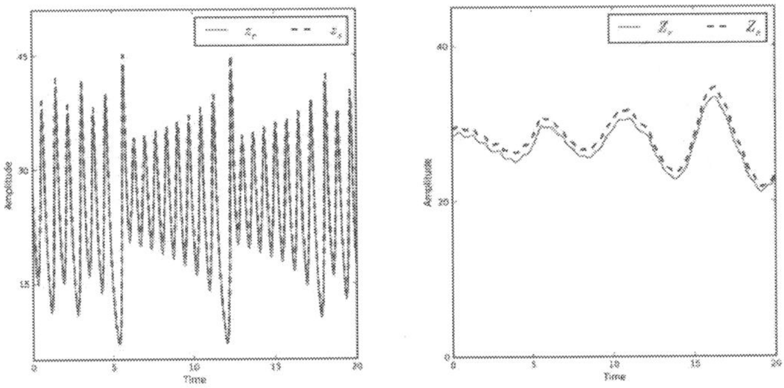


**Figure 4.** Original (in red) and pseudo (in blue) orbits for fast $z$ and slow $Z$ variables for $c = 0.015$.

Here, we consider two sets of numerical experiments: weak coupling ($c = 0.015$) and strong coupling ($c = 0.8$) between fast and slow systems. As an example, the original and pseudo-trajectories for the fast $z$ and slow $Z$ variables are shown in **Figures 4** (weak coupling) and **5** (strong coupling). Pseudo-trajectories were calculated using the LSS method by adding a small perturbation to the parameter $r$. The deviation that is a difference between the 'true' and the

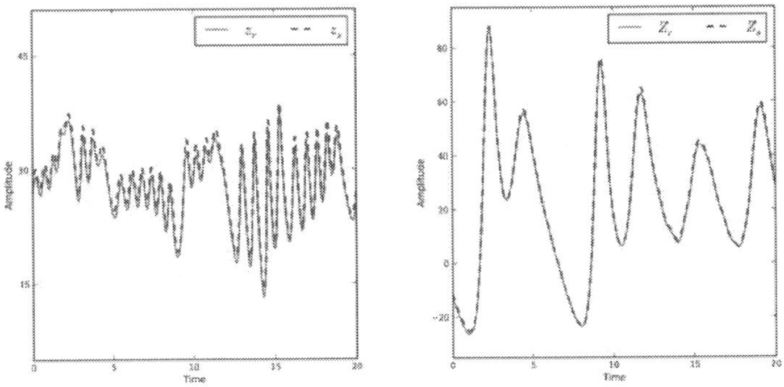


**Figure 5.** Original (in red) and pseudo (in blue) orbits for fast $z$ and slow $Z$ variables for $c = 0.8$.

approximate variables of the slow and fast sub-systems for weak coupling are illustrated in **Figure 6**. Obtained pseudo-trajectories lie very close to the associated true trajectories, proving the existence of shadowing property. Therefore, one can analyse the sensitivity of coupled dynamical system (20) with respect to parameters by averaging the computed sensitivity functions along the trajectory.

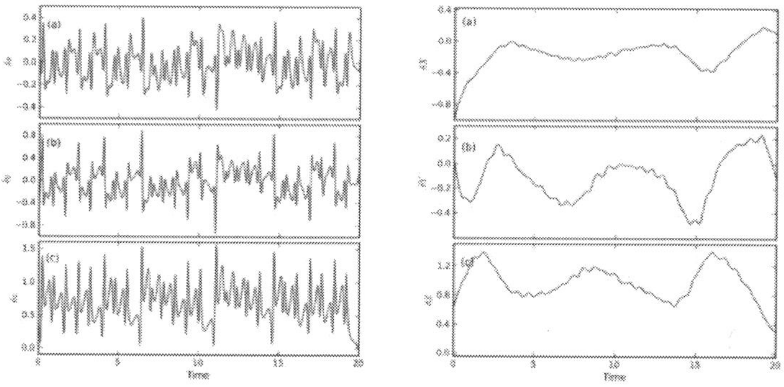

**Figure 6.** Differences between variables that correspond to the original trajectories and pseudo orbits for $c = 0.01$.

| $c$ | $\partial Z/\partial r$ | $\partial Y/\partial r$ | $\partial X/\partial r$ | $\partial z/\partial r$ | $\partial y/\partial r$ | $\partial x/\partial r$ |
|---|---|---|---|---|---|---|
| 1.0 | 1.10 | 0.05 | 0.01 | 1.08 | 0.04 | 0.03 |
| 0.8 | 0.69 | 0.08 | 0.03 | 1.02 | 0.07 | 0.07 |
| 0.4 | 0.95 | 0.03 | −0.01 | 1.03 | 0.09 | 0.09 |
| 0.15 | 0.91 | −0.08 | −0.09 | 1.01 | −0.01 | −0.01 |
| $10^{-4}$ | 1.04 | −0.02 | −0.03 | 1.02 | −0.01 | −0.01 |

**Table 1.** Sensitivity estimates of fast and slow variables with respect to parameter $r$.

The strong coupling does not introduce significant qualitative and quantitative changes in the behaviour of pseudo-trajectories with respect to the true orbits. Sensitivity estimates with respect to the parameter $r$ calculated over the time interval $[0, 20]$ for different values of coupling strength parameter are shown in **Table 1**. The most sensitive variables are $z$ and $Z$. The sensitivity of variables $x$, $y$, $X$ and $Y$ with respect to $r$ is significantly less than that of variables $z$ and $Z$.

The use of shadowing method of sensitivity analysis allows the calculation of average sensitivity functions that can be easily interpreted. Let us recall that sensitivity functions obtained via conventional methods are uninformative and inconclusive because their envelopes grow exponentially over time and functions themselves oscillate.

## 7. Fluctuation dissipation relation in climate sensitivity

To estimate the ensemble-averaged response of the climate system to small external forcing, Leith [38] has proposed using the fluctuation-dissipation theorem (FDT) from statistical physics that quantifies the relation between the fluctuations in a system at thermal equilibrium and the response of the system to applied external perturbations. The FDT was initially obtained by Nyquist [39] in terms of the relation between fluctuations in voltage appearing across a resistor and its impedance. Afterwards, the FDT was formulated and proved for systems modelled by a Langevin equation [40] and for general non-linear dynamical systems in thermodynamic equilibrium [41, 42]. According to the FDT, under certain assumptions, the response of stochastic dynamical system to infinitesimal external perturbations is described by the covariance matrix of the unperturbed system. The two key assumptions commonly considered are: (a) the system state is close to equilibrium, and (b) the probability density function of the unperturbed system is Gaussian. Meanwhile, for a climate system, the standard hypotheses of equilibrium statistical mechanics do not hold since the climate system is highly non-linear and dissipative, and is affected by strong external periodic and stochastic forcing.

In spite of strong non-linearity in the climate system, the linear approximation and time-invariant hypothesis are still extensively applied in climate research [43]. A basic assumption in a linear approximation is that the different external perturbations are acting independently and additively on the system's response. Thus, the response of some climate variable $x$ to external forcing $\Delta\mathcal{F}$ can be represented as $\Delta x = S\,\Delta\mathcal{F}$, where $S$ is a sensitivity function (coefficient). For example, to estimate the change in equilibrium surface temperature $\Delta\Theta_s$ due to the increase in radiative forcing, the following simplified relationship between carbon dioxide and radiative forcing can be used [44, 45]:

$$\cdot\mathcal{F}_{CO_2}(t) = 5.35 \times \ln\left[c(t)/c(0)\right]$$

where $c(t)$ is the concentration of carbon dioxide in parts per million by volume at time $t$ and $c(0)$ is the reference concentration. Therefore, $\mathcal{F}_{CO2}$, where $S = 0.8\ K \cdot W^{-1}m^2$. For doubling of $CO_2$ concentration, this gives the warming of ~ 3 K.

The response of non-linear systems to external perturbations is fundamentally different from the reaction of linear systems [45]. This difference is mainly due to the wider involvement of the system's inherent characteristics and irregularity, and various ways of taking them into account. Consequently, system's fluctuations represent the integration of external forcing and internal feedbacks. The FDT allows us to clearly identify external forcing and separate them from the system's natural oscillations. Let us consider a finite-dimensional dynamical system described by state vector $x$:

$$\dot{x} = f(x),\ x(0) = x_0,\ x \in \mathbb{R}^n \tag{27}$$

Since we are interested in statistical characteristics of this system, let us introduce its average state:

$$\langle x \rangle = \lim_{T \to \infty} \frac{1}{T} \int_0^T x(t)\,dt \tag{28}$$

Together with the system (27), let us consider the perturbed system by adding some external forcing $\delta\mathcal{F}$ to the right-hand side of (27):

$$\dot{x}' = f(x') + \delta\mathcal{F},\ x'(0) = x'_0 \tag{29}$$

If Eqs. (27) and (29) have stationary probability density functions $\rho$ and $\rho'$ respectively, then

$$\langle x \rangle = \int x \rho(x)\,dx,\quad \langle x' \rangle = \int x' \rho'(x')\,dx' \tag{30}$$

It is important to consider that the average state of the perturbed system (29) $\langle x' \rangle$ can be different from that of the unperturbed system (27) and, therefore, the difference $\delta x = \langle x \rangle - \langle x' \rangle$ depends on the external forcing [46]:

$$\langle \delta x \rangle = \mathcal{N}(\delta\mathcal{F})$$

where $\mathcal{N}$ is a certain non-linear function. For small perturbations, we can expect that the relation between $\delta\mathcal{F}$ and $\delta x$ is nearly linear. If $\mathcal{N}$ is differentiable at a certain point $\mathcal{F}_0$, then $\mathcal{N}$ can be represented by a Taylor series. Then if we omit all the non-linear terms and *leave the first order, linear terms*, we obtain:

$$\langle \delta x \rangle = \mathcal{L}(\delta\mathcal{F}) + \text{H.O.T}$$

where $\mathcal{L} = \left.\frac{\partial \mathcal{N}}{\partial(\delta\mathcal{F})}\right|_{\mathcal{F}=\mathcal{F}_0}$ is a linear response operator whose properties are not known *a priori*. Thus, the climate sensitivity problem is to find an operator $\mathcal{L}$. However, this problem is not trivial due to the complicated fractal structure of the attractors (as the set of states) of chaotic systems. Attractors are structurally stable with respect to small external perturbations only for hyperbolic dynamical systems [35, 36], and the response of these systems is linear with respect to small enough external forcing [47]. Since the climate system is not hyperbolic and its attractors depend on external perturbations, some special approaches should be considered. The use of $\epsilon$-regularization of fractal attractor of chaotic dynamical systems allows us to ensure that the attractor ceases to be fractal [48]. The main idea of this method is to add random noise to the right-hand side of the model Eq. (27) [46, 49, 50]:

$$\dot{x} = f(x) + \epsilon\mu(t),\ x(0) = x_0 \tag{31}$$

where $\epsilon$ is a small positive constant and $\mu(t)$ is Gaussian stochastic process. This idea has a reasonable physical base since physical mechanisms that are responsible for energy transformation and dissipation in the climate system are never accurately known. In addition, computer simulation of the system (27) will generate pseudorandom noise due to round-off errors. All of these effects may be considered as white Gaussian noise. Resulting, we can get the Fokker-Plank equation corresponding to (26) for studying the evolution of the probability measure (probability density function) $\rho_\epsilon$ [46, 50]:

$$\frac{\partial \rho_\epsilon}{\partial t} = d\nabla\rho_\epsilon - \mathrm{div}\left[f(x)\rho_\epsilon\right] \tag{32}$$

The function $\rho_\epsilon$ satisfies the following conditions: $\rho_\epsilon \geq 0$ and $\int \rho_\epsilon dx = 1$. If $x \in X$, where $X$ is a compact manifold without a boundary, then (i) the stationary solution of the Eq. (31) exists, (ii) this solution is unique and (iii) this solution is asymptotically stable [46]. In a general case of the phase space $\mathbb{R}^n$, the problem, unfortunately, remains unsolved. In a similar way, we can obtain the Fokker-Plank equation for the perturbed system:

$$\frac{\partial \rho'}{\partial t} = \epsilon^2\nabla\rho' - \mathrm{div}\left[\left(f(x') + \delta\mathcal{F}\right)\rho_\epsilon'\right] \tag{33}$$

The problem now is to find the relation between $\langle\delta x\rangle$ and $\delta\mathcal{F}$. If the increment $\delta\mathcal{F}$ is small enough then

$$\langle\delta x\rangle = \mathcal{L}(\delta\mathcal{F}) = \int_0^t \int x(t+\tau)\left[B(x(t))\right]^{\mathrm{T}} \rho_\epsilon dx d\tau \delta\mathcal{F} \tag{34}$$

where $B = -(1/\rho_\epsilon)\nabla\rho_\epsilon$. If the system dynamic is a stationary random process and $\rho_\epsilon$ is normally distributed, then the response operator can be represented as

$$\mathcal{L}(t) = \int_0^t C(\tau)C(0)^{-1}\,d\tau \tag{35}$$

and we can obtain the following relation between $\langle\delta x\rangle$ and $\delta\mathcal{F}$:

$$\langle\delta x(t)\rangle = \int_0^t C(\tau)C(0)^{-1}\,d\tau\cdot\delta\mathcal{F}(t) \tag{36}$$

where $C(\tau)$ is a $\tau$-lagged covariance matrix of $x$. From this equation, it follows that the response operator for ergodic dynamical system can be calculated from a single trajectory. However, the state vector is dimensionally large; therefore, some procedure should be used to reduce its dimensionality in order to calculate the covariance matrix effectively. For the atmosphere, Eq. (36) was first obtained in [38] under very strict assumptions [42]: (a) the system has at least one quadratic invariant (energy); (b) it has an incompressible phase volume; and (c) it is forced by a weak source. Under these conditions, the system probability density function is Gaussian. Climate systems do not satisfy these assumptions since they have a contracting phase space and do not have any exact quadratic invariant.

The derivation of Eq. (36) represented in [46] is based on more realistic and less restrictive assumptions that make the true validity of Eq. (36) for any non-normal distribution $\rho_\epsilon$. It can be shown that the relation (36) holds under: (a) weak stochastic regularization; (b) weak forcing anomaly and (c) the system has a Fokker-Planck equation with a unique stationary solution.

As a simple example, consider one-dimensional stochastic dynamical system, i.e. system having only one variable $\Theta$ generated by the Langevin equation

$$\dot{\Theta} + \alpha\Theta = \mathcal{F}(t) \tag{37}$$

Here $\alpha = 1/\tau_0$, where $\tau_0$ is the relaxation time for $\Theta$, and $\mathcal{F}(t)$ is a broadband noise forcing such that

$$\langle\mathcal{F}(t)\rangle = 0,\quad \mathcal{F}(t)\mathcal{F}(t') = \vartheta\delta(t-t') \tag{38}$$

Angular brackets denote ensemble averages and $\delta$ is the Dirac delta function. Eq. (37) represents a simple version of radiative balance climate model, which is forced by an external heating whose time dependence is white noise [51, 52]. The variable $\Theta$ is the departure from steady state $\Theta_0$, i.e. $\Theta$ is a climate anomaly. The relaxation time $\tau_0$ in Eq. (36) is $\tau_0 \approx 58$ days [52]. The autocorrelation function for system (37) is

$$\frac{\Theta(t)\Theta(t')}{\Theta^2} = e^{-\alpha|t-t'|}$$

Solving Eq. (37) yields

$$\Theta(t) = \Theta(0)e^{-\alpha t} + \int_0^t e^{-\alpha|t'-t|}\mathcal{F}(t')dt' \tag{39}$$

Since the ensemble average of $\mathcal{F}(t)$ vanishes, one can average Eq. (37) to obtain

$$\langle\dot{\Theta}\rangle + \alpha\langle\Theta\rangle = 0$$

Thus, the average anomaly $\langle\Theta\rangle$ decays exponentially to zero in the relaxation time $\tau_0$. Let the climate system be in equilibrium. Then at $t = t'$, the system is perturbed by a sudden infinitesimal delta function impulse $\vartheta\delta(t - t')$ to the temperature. Consequently, $\Theta(t)$ at $t = t'$ is changed on the value of $\vartheta$. Then the temperature anomaly begins to decrease back to zero in accordance with Eq. (39). The mean system response to any infinitesimal change in the forcing $\mathcal{F}(t)$ can be calculated by the Eq. (37). In particular, for $\mathcal{F}(t)$ being a staircase function, i.e. a constant $\Delta_s$ that is activated at $t = 0$, the system response is given by

$$\langle\Theta\rangle = \Delta_s \lim_{t\to\infty}\int_0^t e^{-\alpha\tau}d\tau = \Delta_s \lim_{t\to\infty}\frac{1-e^{-\alpha t}}{\alpha} = \frac{\Delta_s}{\alpha}$$

The validity of relation (36) was verified for different models of the atmosphere [46]. This relation holds with a high accuracy for both barotropic and two-layer baroclinic global atmospheric models and with a satisfactory accuracy for global general circulation models of the atmosphere.

## Author details

Sergei A. Soldatenko[1*] and Denis Chichkine[2]

*Address all correspondence to: s.soldatenko@bom.gov.au

1 Australian Bureau of Meteorology, Melbourne, Australia

2 Department of Mathematics, University of Toronto, Toronto, Canada

## References

[1] Robinson WA (2001) Modeling Dynamic Climate System. New York: Springer-Verlag, 213 p.

[2] Marshall J, Plumb RA (2007) Atmosphere, Ocean, and Climate dynamics: An Introductory Text. Burlington: Elsevier Academic Press, 344 p.

[3] Dijkstra HA (2013) Nonlinear Climate Dynamics. New York: Cambridge University Press, 367 p.

[4] McGuffie K, Henderson-Sellers A (2014) The Climate Modeling Premier. 4th edition. New York: Wiley-Blackwell, 456 p.

[5] Barry RG, Hall-McKim EA (2014) Essentials of the Earth's Climate System. New York: Cambridge University Press, 271 p.

[6] Goosse H (2015) Climate System Dynamics and Modeling. New York: Cambridge University Press, 378 p.

[7] Pedlosky J (1987) Geophysical Fluid Dynamics. New York: Springer-Verlag, 710 p.

[8] Pedlosky J (2003) Waves in the Ocean and Atmosphere. Berlin-Heidelberg: Springer-Verlag, 264 p.

[9] Katok A, Hasselbatt B (1987) Introduction to the Modern Theory of Dynamical Systems. New York: Cambridge University Press, 802 p.

[10] Tchuenche JM (2013) Dynamical Systems: Theory, Applications & Future Directions. New York: Nova Publ., 312 p.

[11] Brin M, Stuck G (2015) Introduction to Dynamical Systems. New York: Cambridge University Press, 256 p.

[12] Dymnikov VP, Filatov AN (1997) Mathematics of Climate Modeling. Boston: Birkhäuser, 264 p.

[13] Gill M, Childress S (1987) Topics in Geophysical Fluid Dynamics: Atmospheric Dynamics, Dynamo Theory, and Climate Dynamics. New York: Springer-Verlag, 512 p.

[14] Buzzi J (2012) Chaos and Ergodic Theory. In: Meyers RA, ed., Mathematics of Complexity and Dynamical Systems. New York: Springer-Verlag, pp. 63–87.

[15] Lorenz E (1963) Deterministic nonperiodic flow. Journal of the Atmospheric Sciences 20: 130–141.

[16] Kaplan JL, Yorke JA (1979) Chaotic behavior of multidimensional difference equations. In: Peitgen HO and Walter HO, eds., Functional Differential Equations and Approximations of Fixed Points. Lecture Notes in Mathematics. Berlin: Springer-Verlag, pp. 228–237.

[17] Pesin BJ (1977) Characteristic Lyapunov exponents and smooth ergodic theory. *Russian Mathematical Surveys 32:* 55–114.

[18] Palmer TN (2014) Predictability of Weather and Climate: From Theory to Practice. In: Palmer T. and R. Hagedorn, eds., Predictability of Weather and Climate. New York: Cambridge University Press, pp. 1–29.

[19] Stocker T (2011) Introduction to Climate Modelling. Advances in Geophysical and Environmental Mechanics and Mathematics.. Berlin-Heidelberg: Springer-Verlag, 182 p.

[20] Eslami M (1994) Theory of Sensitivity in Dynamical Systems: An Introduction. Berlin-Heidelberg: Springer-Verlag, 601 p.

[21] Rosenwasser E, Yusupov R (2000) Sensitivity of Automatic Control Systems. Boca Raton: CRC Press, 456 p.

[22] Cacuci DG (2003) Sensitivity and Uncertainty Analysis, Volume I: Theory. Boca Raton: CRC Press, 285 p.

[23] Cacuci DG, Ionesku-Bujor M, Navon IM (2005) Sensitivity and Uncertainty Analysis, Volume II: Applications to Large-Scale Systems. Boca Raton: CRC Press, 353 p.

[24] Bofetta G, Crisanti A, Paparella F, Provenzale A, Vulpiani A (1998) Slow and fast dynamics in coupled systems: A time series analysis view. Physica D 116: 301–312.

[25] Peña M, Kalnay E (2004) Separating fast and slow models in coupled chaotic systems. Nonlinear Processes in Geophysics 11: 319–327.

[26] Soldatenko S, Chichkine D (2014) Correlation and spectral properties of a coupled nonlinear dynamical system in the context of numerical weather prediction and climate modeling. Discrete Dynamics in Nature and Society 2014: 498184.

[27] Rosenblum M, Pikovsky A, Kurth J (1996) Phase synchronization of chaotic oscillators. Physical Review Letters 76: 1804–1807.

[28] Liverani C (1995) Decay of correlation. Annals of Mathematics 142: 239–301.

[29] Panchev S, Spassova T (2005) Simple general circulation and climate models with memory. Advances in Atmospheric Sciences 22: 765–769.

[30] Lea D, Allen M, Haine T (2000) Sensitivity analysis of the climate of a chaotic system. Tellus 52A: 523–532.

[31] Wang Q (2013) Forward and adjoint sensitivity computation for chaotic dynamical systems. Journal of Computational Physics 235: 1–13.

[32] Soldatenko S, Steinle P, Tingwell C, Chichkine D (2015) Some aspects of sensitivity analysis in variational data assimilation for coupled dynamical systems. Advances in Meteorology 2015: 1–22.

[33] Pilyugin S (1999) Shadowing in Dynamical Systems, Lecture Notes in Mathematics. Berlin: Springer-Verlag, 276 p.

[34] Palmer HJ (2000) Shadowing in Dynamical Systems. Theory and Applications.. Dordrecht: Kluwer, 299 p.

[35] Anosov DV (1967) Geodesic flows on closed Riemann manifolds with negative curvature. Proceedings of the Steklov Mathematical Institute 90: 1–209.

[36] Bowen R (1975) Equilibrium States and the Ergodic Theory of Anosov Diffeomorphisms. Lecture Notes in Mathematics.. Berlin: Springer-Verlag, 77 p.

[37] Wang Q, Hu R, Blonigan P (2014) Least squares shadowing sensitivity analysis of chaotic limit cycle oscillations. Journal of Computational Physics 267: 210–224.

[38] Leith CE (1975) Climate response and fluctuation dissipation. Journal of the Atmospheric Sciences 32: 2022–2026.

[39] Nyquist H (1928) Thermal adaptation of electric charge in conductors. Physical Review 32: 110–113.

[40] Callen HB, Welton TA (1951) Irreversibility and generalized noise. Physical Review 83: 34–41.

[41] Kubo R (1957) Statistical-mechanical theory of irreversible processes. I. General theory and simple applications to magnetic and conduction problems. Journal of Physical Society of Japan 12: 570–586.

[42] Kraichnan RH (1959) Classical fluctuation-relaxation theorem. Physical Review 113: 1181–1182.

[43] IPCC, 2013: Climate Change 2013: The Physical Science Basis. Contribution of Working Group I to the Fifth Assessment Report of the Intergovernmental Panel on Climate

Change. Stocker TF, Qin D, Plattner GK, Tignor M, Allen SK, Boschung J, Nauels A, Xia Y, Bex V and Midgley PM, eds., Cambridge: Cambridge University Press, 1535 p.

[44] Myhre G, Highwood EJ, Shine KP, Strodal F (1998) New estimates of radiative forcing due to well mixed greenhouse gases. Geophysical Research Letters 25: 2715–2718.

[45] Ruzmaikin A (2014) Climate as a game of chance. Physics Uspekhi 184: 297–311.

[46] Gritsun A, Branstator G (2007) Climate response using a three-dimensional operator based on the fluctuation-dissipation theorem. Journal of the Atmospheric Sciences 64: 2558–2575.

[47] Ruelle D (1998) General linear response formula in statistical mechanics and the fluctuation-dissipation theorem far from equilibrium. Physics Letters A 245: 220–224.

[48] Zeeman EC (1998) Stability of dynamical systems. Nonlinearity 1: 115–135.

[49] Risken H. (1984) The Fokker-Plank Equation: Methods of Solution and Applications. New York: Springer-Verlag, 454 p.

[50] Dymnikov VP (2002) Fluctuation-dissipation relations for dynamic stochastic equations with coefficients periodically dependent on time for dissipative systems with random forcing. Atmospheric and Oceanic Physics 38: 658–663.

[51] Budyko MI (1969) The effect of solar radiation variations on the climate of the earth. Tellus 21: 611–619.

[52] North GR, Cahalan RF (1981) Predictability of a solvable climate model. Journal of the Atmospheric Sciences 38: 504–513.

Chapter 6

# The Eta Model: Design, Use, and Added Value

Fedor Mesinger, Katarina Veljovic, Sin Chan Chou,
Jorge Gomes and André Lyra

Additional information is available at the end of the chapter

http://dx.doi.org/10.5772/64956

**Abstract**

The design of the Eta model goes back to early 1970s, when its original dynamical core was designed following the philosophy of Akio Arakawa of emulating important properties of the atmospheric governing equations. The core's later major features were invented and implemented in the mid-1980s. Once a comprehensive physics package was added, the model became operational as a regional NWP model in the United States in 1993. Its use for regional climate projections followed later, for the South American region and then for a regional reanalysis over the North American region. Summary of the model's dynamical core is given, followed by that of its physics package. Results of experiments revealing the model's ability to generate added value even at large scales when run as a regional climate model (RCM) are summarized. The Eta model is applied on various climate scales seamlessly, from subseasonal, seasonal to multidecadal, from coarse 40 km up to high 5 km resolution. Examples of applications to various socioeconomic sectors, such as for hydropower management, crop yield forecasts, environmental and forest conservation, urban areas management, assessment of natural disaster risks, etc., are given. The Eta RCM capability to reproduce extreme climatic values is pointed out.

**Keywords:** Eta model, eta vertical coordinate, regional climate models, topography in climate models, added value by RCMs, horizontal diffusion

## 1. Introduction: model design

The origins of the Eta model go back as far as 1973 when the dry version code was written at the University of Belgrade of its "ancestor" model, referred to at the time as limited area primitive equation model (LAPEM). This first code already had an Arakawa, or Arakawa-Lorenz, conserving vertical advection scheme, and a lateral boundary condition (LBC) scheme that stayed in place until the present.

Over the years, many development steps followed along with a few model name changes until the name "Eta" became generally accepted, referring to the model´s unique vertical coordinate. But a number of other features that deserve to be noted are found in model's dynamical core, which can be summarized as follows:

- For the gravity wave terms, on the model's E grid, forward-backward scheme is used that

  - avoids the time computational mode of the leapfrog scheme and is neutral with time steps twice leapfrog [1];

  - is modified to enable propagation of a height point perturbation to its nearest-neighbor height points, thereby suppressing the space computational mode of the semistaggered grid [2, 3].

- Also for other terms, including various physics calls, split explicit time differencing is used. This makes the model very efficient since long time steps are used for subroutines that do not require short steps for reasons of stability.

- Horizontal advection scheme is used that conserves energy and C-grid enstrophy, on the B/E grid, in space differencing, for two-dimensional nondivergent flow. The scheme is one of Janjić [4] obtained by transformation of the Arakawa and Lamb [5] C-grid scheme so as to use the B/E grid velocity components.

- Conservation of energy is enforced in transformations between the kinetic and potential energy, in space differencing [6].

- Nonhydrostatic version is available via a switch of the code. If chosen, the model uses the scheme of Janjic [7]. Instead of solving a prognostic equation for the vertical velocity component, the scheme approximates the vertical acceleration using the finite difference of the hydrostatic vertical velocities of two consecutive time steps.

- The eta vertical coordinate [8] ensures hydrostatically consistent calculation of the pressure-gradient ("second") term of the pressure-gradient force (PGF) irrespective of the steepness of the terrain. This is because the eta coordinate surfaces are very nearly horizontal. The topography discretization with the coordinate has been upgraded some years ago by introducing the so-called sloping steps (e.g., [9]), which successfully addresses the Gallus and Klemp [10] problem the step-topography discretization has with flow over a bell-shaped topography [11, 12].

- Van Leer type finite-volume vertical advection of dynamic variables (v, T), using the scheme of [13].

One advantage of the eta coordinate compared to the usual terrain-following (sigma) is that vertical sides of model cells are very nearly equal so that for a finite-volume discretization, which ensures conservations by keeping track of the inflow and outflow into and out of model cells, multiplications of horizontal fluxes by the area of the vertical sides of cells can be skipped.

This is done in the Eta so that in view of the Arakawa-type horizontal advection and finite-volume vertical advection it is very nearly a finite-volume dynamical core model. In fact, in view of its conservation of second-order integral quantities of enstrophy and kinetic energy, not done in standard finite-volume cores, its dynamical core can be referred to as a finite-volume "plus" dynamical core.

The physics package of the model includes the Noah surface scheme over land and sea ice, with roughness length following [14], and Monin-Obukhov with Paulson functions inside the surface layer. In addition, the wind-direction-dependent form drag scheme is used to account for the subgrid scale topography [15]. Over water, surface fluxes scheme is used with a molecular sublayer, roughness length after Charnock, and Mellor-Yamada level 2 derived scheme inside the surface layer. Mellor-Yamada 2.5 turbulence is used above the surface layer, with refinements that include addressing its realizability problem following [16–19]. A choice of the Betts-Miller-Janjic [9, 20] or Kain-Fritsch convection scheme is available. The momentum transport scheme [9, 21] is present within the model's Kain-Fritsch scheme. A choice of scheme [22] or [23] is used for cloud microphysics. GFDL radiation schemes [24, 25] are used, but at the time of this writing, they are being replaced by the RRTMG scheme of [26], upgraded subsequently by [27]; for the impact in clear sky case, see Ref. [28].

The model's lateral boundary condition (LBC) scheme makes an effort to respect the mathematical nature of the problem, prescribing prognostic variables only along the outside boundary grid points, with one less variable prescribed at the outflow than at the inflow points [1]. At outflow points, the velocity component tangential to the boundary is extrapolated from inside of the domain. Experiments of [29] strongly indicate that thereby a result is obtained generally in better agreement with observations than when using the Davies [30] boundary relaxation scheme, used by just about all other limited area models.

## 2. Eta coordinate treatment of topography

After about 2000 quite a few papers stated that the eta coordinate, used by the Eta model, "has difficulties in representing flow over mesoscale topography" and/or that it "appears to be ill suited for high-resolution prediction models" (e.g., [31], among others).What prompted these doubts was a poor result of an experimental NCEP Eta model in forecasting a downslope windstorm in the lee of the Utah Wasatch Range, and even more the results of Gallus and Klemp [10]. In simulating an idealized case of flow over a bell-shaped topography, or "Witch of Agnesi", and using an eta nonhydrostatic code, Gallus and Klemp [10] had flow separate downstream of the mountain instead of properly descending along the lee slope.

To address the problem, discretization using the eta coordinate was refined by assuming sloping bottoms of model's lowest velocity cells, if the four neighboring elevation values permit the definition of the direction of the slope. This would be the case when one of the four elevation values is higher than the remaining three, and when two neighboring values are equal and higher than the remaining two. The slope would be defined inside the model layer below that of the velocity point. For an illustration in the simplest 2D case, see **Figure 2** of Ref. [9].

The refinement improved considerably the simulation of flow over the bell-shaped topography, **Figure 3** of [9], but not as much as the artificial enforcement of zero vorticity along the step corners by Gallus and Klemp, a modification they found to be working better than some others they tried. Recently, however, an oversight was noticed in the sloping steps Eta code, consisting of its horizontal diffusion not having been made aware of the existence of the slopes. While addressing this oversight, in addition the horizontal diffusion scheme was made unconditionally stable and monotonic. This seemed desirable because in very high resolution runs apparently too high values of the scheme's nonlinear diffusion coefficients were resulting from high values of the velocity deformation. Following these refinements, the simulation of the Gallus-Klemp experiment was obtained as shown in the right-hand plot of **Figure 1**. The result that Gallus and Klemp obtained with their artificial code modification is shown in the left-hand plot. The right-hand plot shows velocities along the lee slope even slightly greater than those of the left-hand plot.

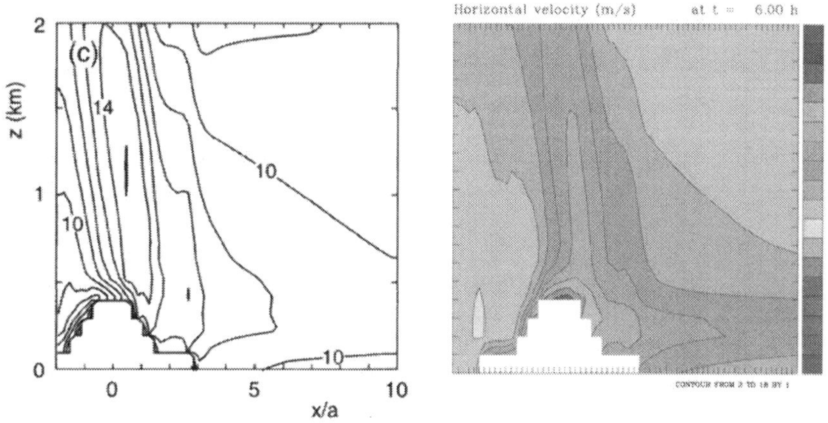


**Figure 1.** Simulation of the Gallus-Klemp experiment with the Eta code allowing for velocities at slopes in the horizontal diffusion scheme, right-hand plot. The plot (c) of Figure 6 of Ref. [10], left-hand plot. Contours show wind speed in m $s^{-1}$, with 10 m $s^{-1}$ prescribed at the lateral boundaries of the domain; see [10] for additional detail.

In summary, with the eta coordinate discretization using sloping steps refinement, and the model's Arakawa-style schemes, arbitrarily steep topography can be used with no significant noise generation and no major disadvantages that we are aware of. One could consider that a downside of the coordinate is that vertical resolution of the boundary layer becomes poor over high topography. However, model layers getting very thin over high topography with terrain-following systems might not be desirable either. For the illustration of the point made about steep topography in **Figure 2** a plot is reproduced from Figure 8 of [9], showing vertical cross section generated using a nonhydrostatic option and running a case of a severe downslope windstorm in the lee of the Andes. Cliffs of well over a kilometer and up to about 2 km are seen in the model's topography, with no visible noise in the isotherms depicted. The plot is

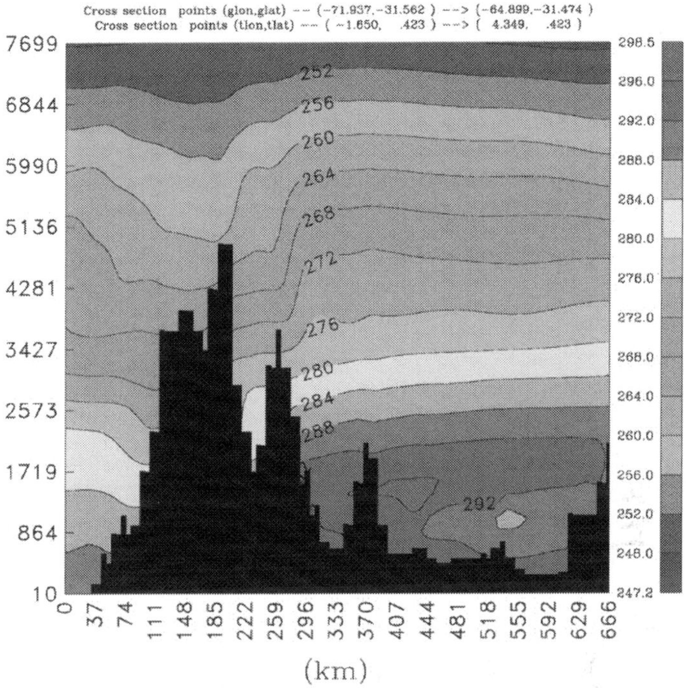


**Figure 2.** West-east cross section of the topography and temperature in deg K at 33 h of an 8 km resolution Eta forecast initialized at 0000 UTC 10 July 2006. The case run is one of an intense zonda windstorm in the lee of the Andes, with heating in the area of the station San Juan, at the distance of about 340 km from the origin of the plot, of more than 9 K compared to the temperature at 24 h. For additional detail see [9].

generated using NCAR Graphics Package with values of individual grid cells shown and no contour smoothing.

## 3. Value added at large scales

One issue the Eta model has been used to address is that of the possibility of achieving added value at large scales inside the domain of a regional climate model (RCM). The usual attitude of people running RCMs is that RCMs should improve on small scales, while keeping the large scales as close as possible to those of the global driver model. To this end, nudging of the RCM's large scales toward those of the driver global model is frequently applied. Opinions have even been advanced that RCMs are unable to improve on large scales. Veljovic et al. [32] have on the other hand argued that attempting to improve also on large scales when running an RCM is a worthwhile objective, and have presented results indicating that this indeed is possible. If one can achieve added value at large scales inside an RCM, small scales will obviously benefit too, and perhaps quite considerably so. Thus, **Figure 1** of [32] shows an example of quite an extraordinary improvement achieved using an RCM, which hardly could have been possible without an added value at large scales.

A caveat is, however, needed regarding some of the tests for which authors have made statements on the added value achieved at different scales but have driven their RCMs by various reanalysis fields. One should notice that it is only via experiments that use GCM results for LBCs that one can fairly test the ability of an RCM to add value at large scales. It is crucial to note that the purpose of RCMs is not to reproduce something we know, but to improve on climate integrations that are projecting changes into the future that we do not know. When driving an RCM with reanalysis LBCs, the RCM has no opportunity to improve on large scales of the climate projection driver data even though it might be able to do so. As pointed out by Veljovic et al. [32], and restated in [9], this is because such climate projection driver data have not been made a part of the experiment. Thus, a proper assessment of the optimal domain size, and of the ability of the RCM to add value at different scales, is not possible.

One should note that we are not saying that in experiments with RCMs driven by a reanalysis one cannot improve on a field or a feature that might be considered large scale, such as perhaps precipitation pattern, compared to the reanalysis used. But this is not a test of the ability of the RCM to add value at large scales because the global fields against which such a test needs to be made, climate projection fields, were not used to drive the RCM.

It is suggested that in order to achieve added value at large scales, four requirements must be observed. First, one needs to run the RCM on a relatively large domain. Note that domain size is very inexpensive as compared to resolution. Second, one should use LBCs that are not ignoring the basic mathematical properties of the problem at hand. That means enforcing the driver model values along the outside RCM boundary only, and not all of them at the outflow parts of the boundary. Third, one must not use large scale or spectral nudging. And finally, fourth, one must use an RCM with dynamical core not of inferior quality to that of the driver global model.

Following the original effort of Veljovic et al. [32], additional experiments that support the four requirements listed above have been presented by Mesinger and Veljovic [11, 12]. For an overall test, the most crucial issue is obviously demonstrating that indeed achieving added value at large scales is possible when the requirements listed are observed. To this end, in **Figure 3** the verification results are shown for an experiment in driving a 21-member Eta model ensemble by ECMWF 32-day ensemble members. Verification scores chosen are for the wind at 250 hPa, assuming that it is the placement accuracy of the wind at upper troposphere, in particular of the jet stream, that reflects the model skill in forecasting atmospheric large scales better or at least more directly than various spectral analysis methods a number of authors used. Therefore, the placement of wind speeds greater than a chosen value, selected at 45 m $s^{-1}$, has been assessed using the verification approach that is customary in quantitative precipitation verifications. ECMWF analyses were taken as "observed". The measure used is the one usually referred to as the equitable threat score (ETS) or Gilbert skill score (e.g., [33]). To minimize the impact of "hedging", e.g., of obtaining a higher score if the forecast area is greater than the observed, the "bias adjusted" ETS scores, ETSa [34], were calculated. Results for the driver

http://dx.doi.org/10.5772/64956

ECMWF ensemble (red) and the Eta ensemble (blue) are shown in the upper panel of the figure. More traditional RMS differences between the ensemble members and the analyses are shown in the lower panel.

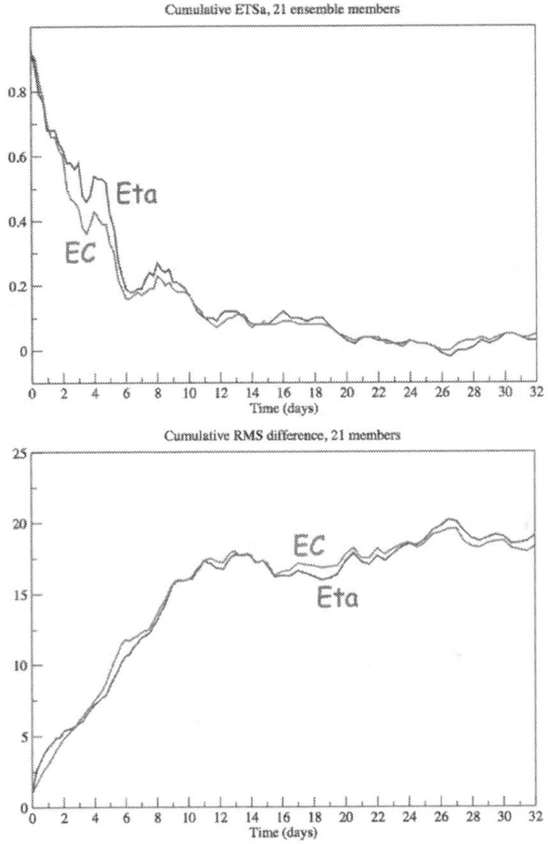


**Figure 3.** Bias-adjusted ETS scores of 250 hPa wind speeds greater than 45 m $s^{-1}$, upper panel, and RMS wind difference, in m $s^{-1}$, lower panel, of the driver ECMWF ensemble members (red) and Eta members (blue), both with respect to ECMWF analyses. Initial time is 0000 UTC 4 October 2012.

Compared to that of the earlier experiment of Veljovic et al. [32], the driver ECMWF ensemble of the experiment of **Figure 3** was of an increased resolution, of about 32 km the first 10 days, and 63 km thereafter. The resolution of the Eta ensemble was unchanged, about 31 km. Thus, during the first 10 days the resolution of the two ensembles of **Figure 3** was about the same. Yet, a clear advantage of the Eta ensemble is seen in the ETSa scores during the first 3–9 days of the experiment, supported by the lower RMS differences during that time. The differences in ETSa scores following day 10 are smaller, with the Eta scores being somewhat higher most of the time when the two are visibly different during about the first half of that period, and an opposite result near the end. The RMS differences following day 10 show about the same message. One might consider scores of earlier times, when the model skill is still synoptically useful, to have more of a meaning than that near the end of the experiment, in particular once the ETSa becomes close to its random value of 0.

Along with the earlier results cited, the scores of **Figure 3** we feel amount to a fairly large body of evidence showing that a nested limited area model, or an RCM, can achieve added value over its driver model also at large scales. If it does so more often than not, then using large scale or spectral nudging will be a nudging in the wrong direction. One should also note that the boundary conditions scheme of Ref. [30] or relaxation boundary conditions, forcing the RCM values in a boundary *band* of points to approach those of the driver model, also represent a way of nudging the RCM toward the large scales of the driver model. Thus, if used, they reduce the ability of the RCM to improve on large scales. In summary, if one finds that spectral or large scale nudging is needed for an acceptable result, we feel it is a good idea to try to find out why.

## 4. Regional reanalysis

Starting with Kalnay et al. [35], a practice of analyzing all reasonably available data using an unchanged system of model and data assimilation has become customary, with the idea that thereby as much as possible the changes in model and data assimilation systems will be taken out of the resulting analyses so that as much as feasible the climate and climate change information only will impact the analyses. Presumably for the first time, in the early two thousands a regional reanalysis effort has been undertaken at NCEP, running the then operational Eta model and data assimilation systems. The paper presenting the project design and selected results for the original 25-year period of 1979–2004 is Mesinger et al. [36]. The reanalysis, named North American Regional Reanalysis (NARR), is continued by NCEP/ Climate Prediction Center in near-real time, and is used for many application purposes.

## 5. From weather to climate

The Eta model has been applied to produce weather forecasts over South America at the Centre for Weather Forecasts and Climate Studies (CPTEC) of INPE since December 1996 [37]. Due to the steep slopes of the Andes Cordillera in South America, the eta coordinate was demonstrated to enable a realistic reproduction of the major features of the summer circulation over South America [38].

The first attempts to extend the integration of the Eta model were applied over South America: Chou et al. [39, 40] using two different land surface schemes—the simple bucket model and the SSiB scheme. The evaluations of one-month simulations produced by the Eta model showed improvement over the driver CPTEC Atmospheric General Circulation Model (AGCM). Chou et al. [40] demonstrated that the model responds most strongly to changes in the lateral boundary conditions, in the second place, to changes in the sea surface temperature, and then, finally, to the land-surface conditions, such as soil moisture and soil temperature.

The Eta model started seasonal forecasts over South America [41] in 2002 by applying persisted anomaly of sea surface temperature as the lower boundary condition of the model. The model was driven by CPTEC AGCM through the lateral boundary conditions. The integration range was 4.5 months. The evaluations showed that the model reproduced the seasonal precipitation variability over the continent and some potential use of higher frequency variability forecast by the model. The Eta seasonal forecasts clearly showed advantage over the driver CPTEC AGCM forecasts, in particular of the seasonal precipitation forecasts [42]. Some systematic errors of seasonal precipitation forecasts were identified in the construction of model seasonal climatology [43] such as the underestimate of 850 hPa specific humidity in the central part of the domain during summer season. Bustamante et al. [43] also showed the interannual variability of seasonal precipitation predicted by the Eta model in agreement with observations. Resende [44] also obtained an underestimate of 700 hPa specific humidity over most of the South America by the Eta model simulations driven by Climate Forecast System Reanalysis (CFSR) [45] at the lateral boundary conditions, in both winter and summer seasons. This underestimate of seasonal-specific humidity may be one of the sources of errors of seasonal precipitation, which is generally underestimated [46].

The relevance of the boundary conditions to drive the regional climate model is shown in Pilotto et al. [47], who compared the Eta seasonal precipitation forecasts driven by CPTEC AGCM, with persisted sea surface temperature anomaly, against the forecasts driven by CPTEC Coupled Ocean-Atmosphere GCM, using predicted sea surface temperature. The model was set up in the domain covering the southern Atlantic Ocean and eastern part of South America. Improvements of the precipitation forecasts over the equatorial oceans by the Eta forecasts driven by the OAGCM were substantial by reducing the excessive precipitation of the Intertropical Convergence Zones over the Atlantic and Pacific Oceans.

A first attempt to apply the Eta RCM forecasts for crop yield productivity in Brazil was produced by Vieira [48] using a physically based crop model driven by Eta seasonal forecasts. Although the evaluation of these forecasts showed some level of error in various areas where corn crop is produced [49], the forecasts captured reasonably well the crop productivity forecasts, particularly in the months of maximum productivity.

Most of the energy production in Brazil is based on hydropower. Therefore, the accurate precipitation forecasts are essential for the correct management of energy production, distribution, and transmission. An example of the application of Eta RCM seasonal forecasts over major Brazilian river basins can be found, for example, in the project mentioned by [50], that demonstrated that the Eta RCM has smaller systematic errors in total precipitation during the rainy season of the Sao Francisco River Basin, in comparison with the driver CPTEC AGCM seasonal precipitation forecasts. These forecasts were run in ensemble mode perturbing parameters of the Betts-Miller-Janjic convection scheme of the Eta model. Another example of the application of the seasonal forecasts for the energy sector is the work of the evaluation of the Eta seasonal forecasts in capturing the onset of rainy period over the Parana River, one of the major rivers for power production in Brazil [51].These forecasts showed large spread in determining the onset of the rainy season. A potential application in the energy sector is the forecast in subseasonal scale as shown in [52] driven by CPTEC Coupled Ocean-Atmosphere General Circulation Model (OAGCM), using lagged ensemble of 20 members constructed from twice daily forecasts during 10 consecutive days.

## 6. Climate change

The Eta model was adapted to run from seasonal to multidecadal range [53] in order to develop capacity for studies of climate change. Some of the changes consisted of allowing seasonal variations of vegetation greenness, sea surface temperature reading off coupled ocean-atmosphere global climate models for any long decadal range, synchronicity with OAGCM calendar, and updated equivalent $CO_2$ concentrations according to the projected future emission scenarios. The model was initially driven by the HadAM3P global atmospheric model and had the present climate simulation evaluated against climatology [53]. The model reproduced reasonably well the South America summer climatological features such as the South Atlantic Convergence Zone and the upper level circulation.

To support impact, vulnerability, and adaptation (IVA) studies for the Brazilian Second National Communication to the UNFCCC, the Eta model was set up at 40 km horizontal resolution, nested in four physics perturbation members of the HadCM3 simulations under the A1B emission scenario [54]. The model was run continuously from 1960 to 1990. The time slice between 1961 and 1990 was considered the reference climate period, while three time slices, 2011–2040, 2041–2070, and 2071–2099, were produced as future climate periods. In order to verify model capability to reproduce large-scale climatic pattern in long-term integrations, Nakićenović and Swart [55] demonstrated that the seasonal mean upper-level winds that were simulated in the 30-year continuous runs agreed with reanalysis winds (**Figure 4**). The Bolivian Anticyclone, which is a major summer feature over South America, was correctly positioned.

For the Brazilian Third National Communication, the strategy to construct the Eta ensemble considered two greenhouse gas emission scenarios, RCP4.5 and RCP8.5 [56], and at least two global climate models, HadGEM2-ES and MIROC5. The runs were divided into four time slices from 1961 to 2005, as the reference climate period, and 2006–2040, 2041–2070, and 2071–2100, as the future climate periods [57, 58]. The horizontal resolution was increased to 20 km. From the Second to the Third National Communication, the Eta model was upgraded following [9]. The major changes were the modification of the vertical advection, making the model a full finite volume, and the refinement of the eta coordinate discretization, which allowed sloping sides of topography cells. This version showed improvement over the previous version, in particular in capturing the downslope windstorms at the lee of the Andes in South America [59], the zonda winds, a foehn type of wind.

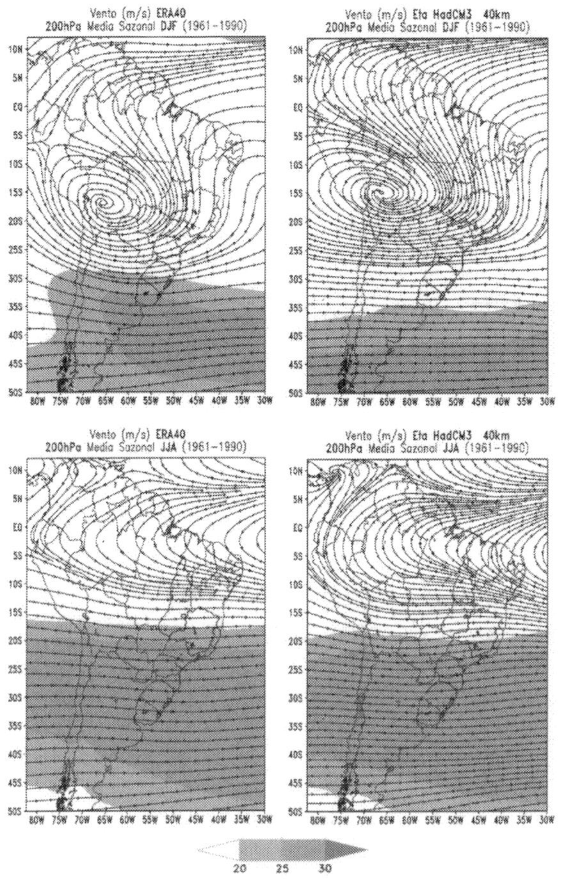


**Figure 4.** Mean 200 hPa streamlines from ERA40 reanalysis (left column) and Eta run nested in HadCM3 simulations (right column) for DJF (top row) and JJA (bottom row). Areas shaded in orange refer to wind speeds in m $s^{-1}$.

The spread of the Eta ensemble runs resulting from the use of different global models at the lateral boundary conditions in [57] was shown to be larger than the spread resulting from the use of the perturbed members of the same global model in [53]. The spread produced by the use of different global model drivers and different emission scenarios [57] attempts to produce the range of lower and upper limits of the projected changes. Therefore, the constructed ensemble tries to envelope the uncertainties associated with the construction of the climate change projections.

The output from the Eta model downscaling of global climate change projections has been applied to support studies on various socioeconomic sectors. Studies of possible impacts of climate change have been produced to the Brazilian hydropower availability [60], to water resources considering small river basins and river springs [61], to Amazon biome and general tropical forest conservation [62, 63], and to coffee crops [64, 65]. Indices of vulnerability and susceptibility to climate change have been designed, for example as in [66], in order to help to construct adaptation measures.

The IVA studies generally require higher resolutions dataset because the problem faced is of local scale rather than global or continental. In addition, the major issue of climate change is the change of the extreme values. In coarse resolution, the values are smoother. The increase in horizontal resolution, easily provided by an RCM, can help to reproduce the frequency distribution of a variable closer to the observed frequency. The evaluations of trends of extreme values in metropolitan areas of São Paulo [67] and Rio de Janeiro [68] under the A1B scenario used the Eta model output at 40 km. The increase in resolution to 20 km [69, 70] has shown improvement over the 40 km, especially in the extreme values. The mean values of the Eta climate change projections at 20 km resolution are very similar to those of the coarse 40 km resolution runs. The gain due to the resolution is detected in the extreme values that agree better with observations as demonstrated by Chou et al. [69].

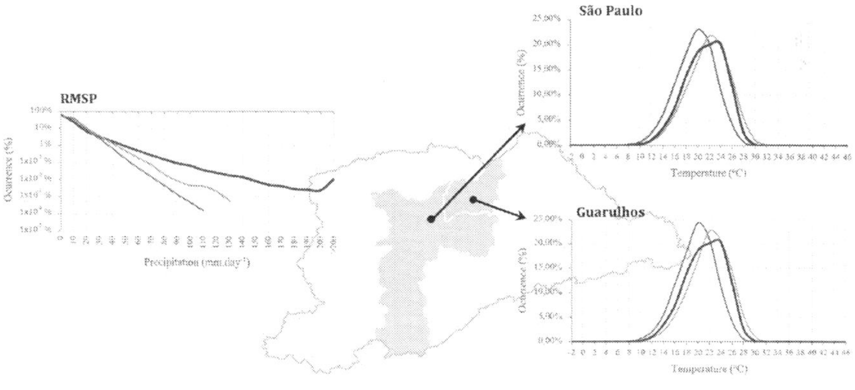


**Figure 5.** Frequency distribution of temperature (°C) for two cities in Brazil, São Paulo and Guarulhos; and frequency distribution of precipitation (*y*-axis in log) (mm day$^{-1}$) for stations around the Metropolitan Region of São Paulo (RMSP, the grey contoured area). The distributions are for the period 1961 to 1990. The black lines refer to observational data; the purple lines, the 20 km Eta simulations; and the blue lines, the 5 km Eta simulations.

An additional horizontal resolution increase may lead to grid sizes smaller than 10 km and to scales in which nonhydrostatic motions become important. The advantage of using the switch to change from hydrostatic to nonhydrostatic mode is clear in long-term integrations, in which resolution is decreased in favor of increasing the integration range. This further increase in horizontal resolution is suitable to help measure the resilience of a city to changes in the climate. The evaluation of the 5 km Eta in nonhydrostatic mode setup over Southeast Brazil [71, 72] showed the advantage of the higher resolution for reproducing the distribution frequency of temperature and precipitation in metropolitan areas. **Figure 5** compares the frequency distribution of temperature and precipitation for stations located in the metropolitan area of

São Paulo obtained by the Eta runs set up in 5 km and in 20 km resolutions. The 5 km resolution distribution approaches the observation distribution more than the 20 km distribution. At this resolution, the topography and coastline are better described, which favor the representation of the winds and temperature, and, consequently, precipitation.

Over southeastern Europe the model was used by Kržič et al. [73] and Djurdjević [74]. Investigations made are similar to some of those summarized above for regions of Brazil. The Eta model used, however, differs in having its radiation scheme replaced by that of [75], and also in being coupled to the Princeton Ocean Model (POM). One needs to note that in these references the acronym EBU is used for the Eta version used, referring to "Eta Belgrade University".

## 7. Work in progress and plans

Along with various applications we are pursuing further development of the model, in several directions. In view of the constant increase in the power of computing resources, one of these efforts is aimed at improving the performance of the model when run at high horizontal resolutions. Thus, we have been making extensive experiments using 1 km resolution over a domain including very complex coastal topography of the Brazilian states of Rio de Janeiro and São Paulo. Our plans addressing the performance at these and higher resolutions include further upgrades of the model's already upgraded Mellor-Yamada 2.5 turbulence scheme. Our codes include some features of the Mellor-Yamada-Nakanishi-Niino scheme (MYNN, e.g., [76], and references therein), and additional work in this area is planned.

Another direction of our model refinement efforts is motivated by the need to enable improved environmental applications at the time of diverse influences on climate by changes in atmospheric constituents, such as aerosols, greenhouse gasses additional to carbon dioxide, etc. Work on coupling these with the now experimentally implemented RRTMG radiation scheme is in progress.

To enable model use in a global setup, work is in progress for implementing the current Eta code to be used mapped on a cubed sphere, by following in the footsteps of the earlier work of Rančić et al. [77] and Zhang and Rančić [78]. Once this work becomes sufficiently mature, it should provide an ideal framework also for the use of the Eta as a nested model on one face of the cube, with resolution higher than on the remaining five faces. This should enable "lateral boundary conditions" of the "nested" model to be defined much more consistently than what can be done when nesting a limited area model in an "alien" global model.

## Acknowledgements

The work of Sin Chan Chou, Jorge Gomes, and André Lyra reported on here was funded by Grant Nos. 400792/2012-5 and 308035/2013-5 of the National Council for Scientific and

Technological Development (CNPq), Brazil. The work of Fedor Mesinger was partially funded by the former of these two grants and by Grant No. F-147 of the Serbian Academy of Sciences and Arts, and that of Katarina Veljovic by the Ministry of Science and Technological Development of the Republic of Serbia under Grant No. 176013.

## Author details

Fedor Mesinger[1,2*], Katarina Veljovic[3], Sin Chan Chou[2], Jorge Gomes[2] and André Lyra[2]

*Address all correspondence to: fedor.mesinger@gmail.com

1 Serbian Academy of Sciences and Arts, Belgrade, Serbia

2 National Institute for Space Research (INPE), Cachoeira Paulista, SP, Brazil

3 Institute of Meteorology, Faculty of Physics, University of Belgrade, Belgrade, Serbia

## References

[1] Mesinger F. Forward-backward scheme, and its use in a limited area model. Contrib. Atmos. Phys. 1977;50:200–210.

[2] Mesinger F. A method for construction of second-order accuracy difference schemes permitting no false two-grid-interval wave in the height field. Tellus. 1973;25:444–458. DOI: 10.1111/j.2153-3490.1973.tb00629.x

[3] Mesinger F, Popovic J. Forward–backward scheme on the B/E grid modified to suppress lattice separation: The two versions, and any impact of the choice made? Meteorol. Atmos. Phys. 2010;108:1–8. DOI: 10.1007/s00703-010-0080-1

[4] Janjić ZI. Nonlinear advection schemes and energy cascade on semi-staggered grids. Mon. Wea. Rev. 1984;112:1234–1245. DOI: 10.1175/1520-0493(1984)112<1234:NASAEC>2.0.CO;2

[5] Arakawa A, Lamb VR. Computational design of the basic dynamical processes of the UCLA general circulation model. Methods Comput. Phys. 1977;17:173–265.

[6] Janjic ZI, Mesinger F, Black TL. The pressure advection term and additive splitting in split-explicit models. Quart. J. Roy. Meteorol. Soc. 1995;121:953–957. DOI: 10.1002/qj.49712152413

[7] Janjic ZI, Gerrity JP Sr., Nickovic S. An alternative approach to nonhydrostatic modeling. Mon. Wea. Rev. 2001;129:1164–1178. DOI: 10.1175/1520-0493(2001)129 <1164:AAATNM=2.0.CO;2

[8] Mesinger F. A blocking technique for representation of mountains in atmospheric models. Riv. Meteorol. Aeronautica. 1984;44:195–202.

[9] Mesinger F, Chou SC, Gomes J, Jovic D, Bastos P, Bustamante JF, Lazic L, Lyra AA, Morelli S, Ristic I, Veljovic K. An upgraded version of the Eta model. Meteorol. Atmos. Phys. 2012;116:63–79. DOI: 10.1007/s00703-012-0182-z

[10] Gallus WA Jr., Klemp JB. Behavior of flow over step orography. Mon. Wea. Rev. 2000;128:1153–1164. DOI: 10.1175/1520-0493(2000)128<1153:BOFOSO>2.0.CO;2

[11] Mesinger F, Veljovic K. Eta vs sigma: Placement of storms, Gallus-Klemp test, and 250 hPa wind skill compared to ECMWF in ensemble experiments. Presentation at NCEP/EMC, Camp Springs, MD, 2015. Available from: http://www.emc.ncep.noaa.gov/seminars/presentations/2015/fm.kv.emc2015.pdf [Accessed: 16-03-08].

[12] Mesinger F, Veljovic K. Eta vs sigma: Review of past results, Gallus-Klemp test, and large scale wind skill in ensemble experiments. Meteorol. Atmos. Phys. 2016; in review.

[13] Mesinger F, Jovic D. The Eta slope adjustment: Contender for an optimal steepening in a piecewise-linear advection scheme? Comparison tests. 2002 NCEP Office Note 439, 29 pp. Available from: http://www.emc.ncep.noaa.gov/officenotes/newernotes/on439.pdf [Accessed: 16-03-08].

[14] Zilitinkevich SS. Non-local turbulent transport: Pollution dispersion aspects of coherent structure of convective flows. In: Power H, Moussiopoulos N, Brebbia CA, editors. Air Pollution III—Volume I: Air Pollution Theory and Simulation. Computational Mechanics Publications: Southampton, Boston, MA; 1995, p. 53–60.

[15] Mesinger F, Wobus RL, Baldwin ME. Parameterization of form drag in the Eta model at the National Centers for Environmental Prediction. In: 11th Conference on Numerical Weather Prediction. American Meteorological Society: Norfolk, VA; 19–23 August 1996, pp. 324–326.

[16] Mesinger F. Forecasting upper tropospheric turbulence within the framework of the Mellor-Yamada 2.5 closure. Research Activities in Atmospheric and Oceanic Modelling, CAS/JSC WGNE Rep. WMO: Geneva; 1993; Vol. 18, pp. 28–29.

[17] Mesinger F. Sensitivity of the definition of a cold front to the parameterization of turbulent fluxes in the NMC's Eta Model. Research Activities in Atmospheric and Oceanic Modelling, CAS/JSC WGNE Rep. WMO: Geneva; 1993; Vol. 18, pp. 36–38.

[18] Janjic ZI. Nonsingular implementation of the Mellor-Yamada Level 2.5 Scheme in the NCEP Meso model. 2002 NCEP Office Note 437, 61 pp. Available from: http://www.emc.ncep.noaa.gov/officenotes/newernotes/on437.pdf [Accessed: 16-03-08].

[19] Mesinger F. Several PBL parameterization lessons arrived at running an NWP model. International Conference on Planetary Boundary Layer and Climate Change, IOP Conf. Series: Earth and Environmental Science 13. IOP Publishing, Bristol, UK; 2010. DOI: 10.1088/1755-1315/13/1/012005

[20] Janjic ZI. The step-mountain eta coordinate model: Further developments of the convection, viscous sublayer, and turbulence closure schemes. Mon. Wea. Rev. 1994;122:927–945. DOI: 10.1175/1520-0493(1994)122<0927:TSMECM>2.0.CO;2

[21] Bastos PR. Inclusion of the momentum perturbation in the Kain-Fritsch cumulus parameterization scheme and the impacts on a convective rain event [MSc dissertation], Instituto Nacional de Pesquisas Espaciais, São José dos Campos, SP, Brazil; 2007.

[22] Ferrier BS, Jin Y, Lin Y, Black T, Rogers E, DiMego G. Implementation of a new grid-scale cloud and precipitation scheme in the NCEP Eta model. In: 19th Conference on Weather Analysis and Forecasting/15th Conference on Numerical Weather Prediction, American Meteorological Society, San Antonio, TX; 12–16 August 2002; pp. 280–283.

[23] Zhao Q, Black TL, Baldwin ME. Implementation of the cloud prediction scheme in the Eta model at NCEP. Wea. Forecasting. 1997;12:697–712. DOI: 10.1175/15200434(1997)012<0697:IOTCPS>2.0.CO;2

[24] Lacis AA, Hansen JE. A parameterization of the absorption of solar radiation in the earth's atmosphere. J. Atmos. Sci. 1974;31:118–133. DOI: 10.1175/1520-0469(1974)031<0118:APFTAO>2.0.CO;2

[25] Schwarzkopf MD, Fels SB. The simplified exchange method revisited: An accurate, rapid method for computation of infrared cooling rates and fluxes. J. Geophys. Res. 1991;96:9075–9096. DOI: 10.1029/89JD01598

[26] Mlawer EJ, Taubman SJ, Brown PD, Iacono MJ, Clough SA. Radiative transfer for inhomogeneous atmospheres: RRTM, a validated correlated-k model for the longwave. J. Geophys. Res. 1997;102:16.663–16.682. DOI: 10.1029/97JD00237

[27] Mlawer EJ, Payne VH, Moncet J-L, Delamere JS, Alvarado MJ, Tobin DC. Development and recent evaluation of the MT_CKD model of continuum absorption. Philos. T. Roy. Soc. A. 2012;370:2520–2556. DOI: 10.1098/rsta.2011.0295

[28] Campos DA, Chou SC, Spyrou C, Chagas JCS, Bottino MJ. Eta model simulations using two radiation schemes in clear sky conditions. Meteorol. Atmos. Phys. 2016, in review.

[29] Mesinger F, Veljovic K. Limited area NWP and regional climate modeling: A test of the relaxation vs Eta lateral boundary conditions. Meteorol. Atmos. Phys. 2013;119:1–16.DOI: 10.1007/s00703-012-0217-5

[30] Davies HC. A lateral boundary formulation for multi-level prediction models. Quart. J. Roy. Meteorol. Soc. 1976;102:405–418. DOI: 10.1002/qj.49710243210

[31] Schär C, Leuenberger D, Fuhrer O, Lüthi D, Girard C. A new terrain-following vertical coordinate formulation for atmospheric prediction models. Mon. Wea. Rev. 2002;130:2459–2480. DOI: 10.1175/1520-0493(2002)130,2459:ANTFVC.2.0.CO;2

[32] Veljovic K, Rajkovic B, Fennessy MJ, Altshuler EL, Mesinger F. Regional climate modeling: Should one attempt improving on the large scales? Lateral boundary

condition scheme: Any impact? Meteorol. Z. 2010;19:237–246. DOI: 10.1127/0941-2948/2010/0460

[33] Hogan RJ, Ferro CAT, Jolliffe IT, Stephenson DB. Equitability revisited: Why the "Equitable Threat Score" is not equitable. Wea. Forecasting. 2010;25:710–726. DOI: 10.1175/2009WAF2222350.1

[34] Mesinger F. Bias adjusted precipitation threat scores. Adv. Geosci. 2008;16:137–143.

[35] Kalnay E, et al. The NCEP/NCAR 40-year reanalysis project. Bull. Am. Meteorol. Soc. 1996;77:437–471. DOI: 10.1175/1520-0477(1996)077<0437:TNYRP>2.0.CO;2

[36] Mesinger F, DiMego G, Kalnay E, Mitchell K, Shafran PC, Ebisuzaki W, Jovic D, Woollen J, Rogers E, Berbery EH, Ek MB, Fan Y, Grumbine R, Higgins W, Li H, Lin Y, Manikin G, Parrish D, Shi W. North American regional reanalysis. Bull. Am. Meteorol. Soc. 2006;87:343–360. DOI: 10.1175/BAMS-87-3-343

[37] Chou SC. Regional Eta Model. Climanálise (Instituto Nacional de Pesquisas Espaciais, São José dos Campos, SP, Brazil). 1996: Special Issue. 203–207.

[38] Figueroa SN, Satyamurty P, Dias PLS. Simulations of the summer circulation over South American region with an eta coordinate model. J. Atmos. Sci. 1995;52:1573–1584. DOI: 10.1175/1520-0469(1995)052<1573:SOTSCO=2.0.CO;2.

[39] Chou SC, Nunes AMB, Cavalcanti IFA. Extended range forecasts over South America using the regional Eta model. J. Geophys. Res. 2000;105:10147–10160.

[40] Chou SC, Tanajura CAS, Xue Y, Nobre CA. Validation of the coupled Eta/SSiB model over South America. J. Geophys. Res. 2002;107(D20):LBA56-1-LBA56-20. DOI: 10.1029/2000JD000270

[41] Chou SC, Bustamante J, Gomes JL. Evaluation of Eta model seasonal precipitation forecasts over South America. Nonlinear Proc. Geophys. 2005;12:537–555. DOI: 10.5194/npg-12-537-2005

[42] Alves LM, Chou SC, Marengo JA. Evaluation of the seasonal forecasts over Brazil from the Eta climate model. In: XIII Congresso Brasileiro de Meteorologia, Fortaleza; September 2004, Brazilian Meteorological Society; 2004.

[43] Bustamante J, Gomes JL, Chou SC. 5-year Eta model seasonal forecast climatology over South America. In: 8th International Conference on Southern Hemisphere Meteorology and Oceanography, Foz do Iguaçu; 24–28 April 2006.

[44] Resende N. Influência das condições do solo na previsão sazonal de extremos climáticos. Masters Dissertation. Instituto Nacional de Pesquisas Espaciais, São José dos Campos, São Paulo, Brazil; 2014.

[45] Saha S, Moorthi S, Pan HL, Wu X, Wang J, Nadiga S, et al. The NCEP climate forecast system reanalysis. Bull. Am. Meteorol. Soc. 2010;91:1015–1057.

[46] Resende N, Chou SC. Influence of soil condition in the climatology of the Eta model seasonal forecasts. Rev. Brasileira Climatol. 2015;15:64–79.

[47] Pilotto ID, Chou SC, Nobre P. Seasonal climate hindcasts with Eta model nested in CPTEC coupled ocean-atmosphere general circulation model. Theor. Appl. Climatol. 2012;110:437–456. DOI: 10.1007/s00704-012-0633-y

[48] Vieira PA. Forecast of climatic features and corn grain yield in the Brazilian Center-South region. PhD Thesis. In Portuguese. Universidade de São Paulo, Esalq, Piracicaba, SP, Brazil; 2006.

[49] Vieira PA Jr., Dourado-Neto D, Chou SC, Martin TN. Eta model forecasts as input to crop models for the Mid-Southern region of Brazil. Ciência Rural. 2009;39:412–420. DOI: 10.1590/S0103- 84782008005000076

[50] Marengo JA, Sampaio G, Chou SC. Seasonal rainfall forecasts doe applications in streamflow generation in the Sao Francisco River Basin in Northeast Brazil: Rainfall corrections for ACGMS for use in hydrological modeling. In: VAMOS! Newsletter. 2006, Department of Atmospheric Science, University of Buenos Aires, Buenos Aires, Argentina; 2006; 3:7–11. Higgins W, Marengo J, editors; 2006.

[51] Chou SC, Bustamante JF, Gomes JL, Tavares P. Forecast of the rainy season onset in the Parana river basin using the Eta model. In: II Simpósio de Recursos Hídricos do Sul-Sudeste. Rio de Janeiro; 2008, Brazilian Water Resources Society; 2008.

[52] Chou SC, Sueiro G, Alves LF, Rodriguez D, Gomes JL, Rozante JR, Nobre P. Precipitation forecasts for 50-day range. In: V Congresso Brasileiro de Energia, Rio de Janeiro; 2015. Energy Planning Program. Federal University of Rio de Janeiro; 2015.

[53] Pesquero JF, Chou SC, Nobre CA, Marengo JA. Climate downscaling over South America for 1961–1970 using the Eta model. Theor. Appl. Climatol. 2009;99:75–93. DOI: 10.1007/s00704-009-0123-z

[54] Nakićenović N, Swart R, editors. Special Report on Emissions Scenarios. A Special Report of Working Group III of the Intergovernmental Panel on Climate Change. Cambridge University Press: Cambridge, United Kingdom and New York; 2000, 599 pp.

[55] Chou SC, Marengo JA, Lyra A, Sueiro G, Pesquero J, Alves LM, Kay G, Betts R, Chagas D, Gomes JL, Bustamante J, Tavares P. Downscaling of South America present climate driven by 4-member HadCM3 runs. Clim. Dyn. 2012;38:635–653.

[56] Stocker TF, Qin D, Plattner G-K, Tignor M, Allen SK, Boschung J, Nauels A, Xia Y, Bex V, Midgley PM, editors. Climate Change 2013: The Physical Science Basis. Contribution of Working Group I to the Fifth Assessment Report of the Intergovernmental Panel on Climate Change. Cambridge University Press: Cambridge, United Kingdom and New York, NY, USA; 2013, 1535 p. DOI: 10.1017/CBO9781107415324

[57] Chou SC, Lyra A, Mourão C, Dereczynski C, Pilotto I, Gomes J, Bustamante J, Tavares P, Silva A, Rodrigues D, Campos D, Chagas D, Sueiro G, Siqueira G, Nobre P, Marengo

J. Evaluation of the Eta simulations nested in three global climate models. Am. J. Clim. Change. 2014;3:438–454. DOI: 10.4236/ajcc.2014.35039

[58] Chou SC, Lyra A, Mourão C, Dereczynski C, Pilotto I, Gomes J, Bustamante J, Tavares P, Silva A, Rodrigues D, Campos D, Chagas D, Sueiro G, Siqueira G, Marengo J. Assessment of climate change over South America under RCP 4.5 and 8.5 downscaling scenarios. Am. J. Clim. Change. 2014;3:512–525. DOI: 10.4236/ajcc.2014.35043

[59] Antico PL, Chou SC, Mourão C. Zonda downslope winds in the central Andes of South America in a 20-year climate simulation with the Eta model. Theor. Appl. Climatol. 2015;123:1–9. DOI: 10.1007/s00704-015-1709-2

[60] Lima JM. The effects of climate change on the generation of electric power. AES Tietê. Hunter Books Editora: São Paulo, Brazil; 2014, 423 pp. ISBN 978-85-68717-00-4.

[61] Viola MR, De Mello CR, Chou SC, Yanagi SN, Gomes JL. Assessing climate change impacts on Upper Grande River Basin hydrology, Southeast Brazil. Int. J. Climatol. 2015;35:1054–1068. DOI: 10.1002/joc.4038

[62] Lyra AA, Chou SC, Sampaio GO. Sensitivity of the Amazon biome to high-resolution climate change projections. Acta Amazonica. 2016;46:175–188. DOI: 10.1590/1809-4392201502225

[63] Lyra AA, Imbach P, Rodriguez D, Chou SC, Georgiou S, Garofolo L. Projections of climate change impacts on central America tropical rainforest. Climatic Change. 2016, in review.

[64] Rodrigues DC, Tavares P, Giarolla A, Chou SC, Resende NC, De Camargo MBP. Estimation of the occurrence of maximum temperature above 34°C during the flowering and maturation of coffee crop based on the Eta/CPTEC 40 km A1B scenario. In: XVII Congresso Brasileiro de Agrometeorologia, Guarapari-ES; July 2011.

[65] Resende NC, Giarolla A, Rodrigues DC, Tavares P, Chou SC. Occurrence of coffee rust disease (Hemileia vastatrix) in some regions of São Paulo, based on the Eta/CPTEC (A1B IPCC/SRES scenario) climatic projections. In: XVII Congresso Brasileiro de Agrometeorologia, Guarapari-ES; July 2011.

[66] Chang MY, Dereczynski C, Freitas MAV, Chou SC. Climate change index: A proposed methodology for assessing susceptibility to future climatic extremes. Am. J. Clim. Change. 2014;3:326–337. DOI: 10.4236/ajcc.2014.33029

[67] Marengo JA, Valverde MC, Obregon GO. Observed and projected changes in rainfall extremes in the Metropolitan Area of São Paulo. Climate Res. 2013;57:61–72.

[68] Dereczynski C, Silva WL, Marengo J. Detection and projections of climate change in Rio de Janeiro, Brazil. Am. J. Climate Change. 2013;02:25–33.

[69] Chou SC, Lyra A, Gomes JL, Marengo JA, Alves L. Future regional climate projection using Eta model. In: Marangon, J, Collischonn W, Marengo J., editors. (Org.). The effects

of climate change on the generation of electric power. 1st ed. Hunter Books Editora: São Paulo, Brazil; 2014, pp. 65–94.

[70] Mourão C, Chou SC, Marengo J. Downscaling climate projections over La Plata Basin. Atmos. Climate Sci. 2016;6:1–12. DOI: 10.4236/acs.2016.61001

[71] Chou SC, Lyra AA, da Silva AJ, Sueiro G, Tavares P, Nunes LH, Marengo JA. 5-km resolution Eta model downscaling of present climate in the City of Santos, Brazil. In: Water Management in Transition Countries as Impacted by Climate Change and Other Global Changes, Lessons from Paleoclimate, and Regional Issues. Institute for the Development of Water Resources and Serbian Academy of Sciences: Belgrade; 3–5 September 2014, Vol. 1, pp. 80–85.

[72] Lyra AA, Tavares P, Chou SC, Sueiro G, Dereczynski C, Sondermann M, Silva A, Marengo J, Giarolla A. Climate change projections over three metropolitan regions in Southeast Brazil using the non-hydrostatic Eta regional climate model at 5-km resolution. Theor. Appl. Climatol. 2016, in review.

[73] Kržič A, Tošić I, Djurdjević V, Veljović K, Rajković B. Changes in climate indices for Serbia according to the SRES-A1B and SRES-A2 scenarios. Climate Res. 2012;49:73–86. DOI: 10.3354/cr01008

[74] Djurdjevic V. Climate change projections international and national level. 2015. Available from: http://www.klimatskepromene.rs/uploads/useruploads/Documents/001-Projekcije-klimatksih-promena_V.Djurdjevic.pdf [Accessed: 2016-03-09].

[75] Chou M-D, Suarez MJ. A solar radiation parameterization for atmospheric studies. NASA Tech. Rep. NASA/TM-1999-10460 1999;15:38.

[76] Nakanishi M, Niino H. Development of an improved turbulence closure model for the atmospheric boundary layer. J. Meteorol. Soc. Japan. 2009;87:895–912. DOI: 10.2151/jmsj.87.895

[77] Rančić M, Purser RJ, Mesinger F. A global shallow-water model using an expanded spherical cube: Gnomonic versus conformal coordinates. Quart. J. Roy. Meteorol. Soc. 1996;122:959–982. DOI: 10.1002/qj.49712253209

[78] Zhang H, Rančić M. A global Eta model on quasi-uniform grids. Quart. J. Roy. Meteor. Soc. 2007;133:517–528. DOI: 10.1002/qj.17